I0471915

I am the true vine, and my father is the vinedresser. Every branch in me that does not bear fruit he takes away: and every branch that does bear fruit, he prunes, that it may bear more fruit

Jn 15:2 ESV

DPs Dominion
Publishing Services
http://www.dominionpublishingstores.yolasite.com

FRUITS AND VEGETABLE TECHNOLOGIES
Handling, processing & storage

Segun R. Bello

B. Eng (Hons), FUT, Akure,   MSc, Ibadan,
MNSE, MNIAE, FSINRHD, R. Engr. (COREN)

Fruit and vegetable technologies
Handling, processing & storage

Federal College of Agriculture Ishiagu, 480001 Nigeria

segemi2002@gmail.com; bellraph95@yahoo.com
http://www.dominionpublishingstores.yolasite.com
http://www.segzybrap.web.com
+234 8068576763, +234 8062432694

ISBN-13: 978- 149-047-910-1
       -10: 149-047-910-4

First Edition published in June 2013

Printed by Createspace US
7290 Investment Drive
Suite B North Charleston,
SC 29418 USA, www.createspace.com

Dedication

This work is dedicated to

Moji, the love of my life

Acknowledgement

I specially acknowledge the hand of God in the pursuit of my career. He gave strength to the meek and humble to do exploit. I express deep appreciations to students of the Department of Horticulture & Landscape Technology and the department of Agricultural Engineering Technology, Federal Colleges of Agriculture Ishiagu for their contributions during interactive class sessions and field discussions which has formed a major resource material in packaging this book and everyone world over who had came in contact with my work and has benefited from the materials in several ways..

The contributions and inputs of Engr. Ezebuilo C. N., Okechukwu of the Department of Agricultural Engineering Technology, Federal College of Agriculture, Ishiagu; Mr. Balogun R. B. of the Department, of Horticultural and Landscape Technology, Federal College of Agriculture, Ishiagu,; Engr. Adegbulugbe T. A. and Femi D. Aremu, of the Department, of Agricultural Engineering Technology, Federal College of Agriculture, Moor Plantation, Ibadan,; Engr. Odey Simon O., of the Agronomy Department, Cross River University of Technology, Obubra campus, Cross River State and other professional colleagues in associated institutions are immensely appreciated.

Despite all the help received from many people, it seems inevitable that there will be some inaccuracies or errors in the text. For these the author accepts responsibility and apologizes in advance for any incorrect statements or impressions given. Should errors be noticed, the author would welcome factual corrections. He would also be happy to receive comments, observations and additional information on any topic, section or statements in any part of the book. This would be particularly useful should any updated or translated edition be planned. Correspondence may be addressed to the author

My sincere appreciation goes to all who at one point or the other, share my visions and mission. Your encouragements, unflinching supports and faithfulness will forever be acknowledged. The nudging from Ayomikun, 'Pelumi, Damilola, Adeola and Bukky are great impetus for the realization of this vision.

Content

Preface xiii

INTRODUCTION xvi

PART 1   CONCEPTS OF FRUITS & VEGETABLE HANDLING xix

CHAPTER 1   Properties of Fruits and Vegetable 3

1.   Introduction 3

1.1   Fruits 3

1.2   Type fruits 4

1.2.1   Fleshy or wet fruits 4

1.2.2   Dry fruits 17

1.2.3   Dried fruits 19

1.3   Tropical fruits 21

1.4   Vegetables 22

1.5   Importance of fruits and vegetables 27

1.6   Nutritional content of fruits and vegetables 27

1.7   Comparison between fruit and vegetable 28

1.8   Fruit development and growth regulator 29

1.8.1   Relevant engineering aspects of growth regulators 29

1.9   Fruit market 33

CHAPTER 2   Fruits and Vegetable Harvest Technologies 34

2   Introduction 34

2.1   Factors affecting changes in fruits and vegetable 34

2.2   Maturity in fruits and vegetable 37

2.2.1   Maturity index and measurement 38

2.3   Ripeness 41

2.4   Fruits and vegetable harvest technologies 44

2.5   Time of harvest factors 44

2.6   Harvest time (picking time) and harvest handling 49

2.7   Fruit harvest technologies 49

2.7.1   Hand picking technology 49

2.7.2   Mechanical method of fruit and vegetable harvest 52

2.8   Harvest effect on postharvest quality 54

2.9   Mechanical damage 54

2.10   Chilling injury 58

Chapter 3   Fruits & Vegetable Quality Measurement 61

3.   Introduction..61

3.1   Meaning of quality...61

3.2   Quality needs of fruit and vegetable 62

3.3   Quality change prediction and maintenance in fruit and vegetable..........62

3.4   Fruit quality indicators 63

3.5   Methods for determining quality of fresh commodities 63

3.6   Factors affecting quality preservation of fresh commodities 68

3.7   Quality measurement of fresh commodities 68

3.7.1   Destructive methods...69

3.7.2   Nondestructive methods...73

3.8   Fruit firmness and its measurement..75

10.1   Fruit firmness instruments 77

10.2   Measuring soluble solids content (ssc) in fruits 80

10.3   Fruit titratable acidity...81

10.4   Maintenance of quality/quality assurance 81

10.5   Types of products contaminants..84

10.6   Cosmetic appearance 84

PART 2   POSTHARVEST SYSTEMS MANAGEMENT & TECHNOLOGIES............85

CHAPTER 4   Postharvest Management Options 86

4.   Introduction...86

4.1   Postharvest practices...87

4.1.1   Transfer from field bin to reception 87

4.1.2   Product cooling at reception...88

4.1.3   Sorting & grading 89

4.1.4   Packaging & labeling 94

4.1.5   Quality control checks..95

4.2   Postharvest management options 95

CHAPTER 5   Postharvest Management Systems in the Tropics 106

5.   Introduction...106

5.1   Postharvest changes in fruits and vegetables...106

5.2   Crop losses in the tropics...107

5.3   Postharvest challenges in the tropics 108

5.5   Challenges to postharvest technology development 112

5.6   Solutions to challenges of postharvest technology development..............114

5.7   The role of agricultural engineers in postharvest development................115

5.8   Future prospects and opportunities in postharvest development.............115

CHAPTER 6   Products Deterioration and Storage Systems 117

6.   Introduction 117

6.1   Factors affecting of products deterioration 117

6.2   Factors affecting stored products deterioration 120

6.2.1. Influence of temperature 121

6.2.2. Influence of relative humidity 122

6.3.   Postharvest storage systems 123

6.4.   Postharvest storage conditions for selected fruits and vegetables 125

6.5.   Fruits and vegetable packaging 129

6.5.1. Properties of packaging materials 129

6.5.2. Fruits and vegetable packaging materials 131

6.6.   Effect of storage and packaging on food 141

6.7.   Storage of horticultural crops 141

6.7.1. Storage practices 142

6.7.2. Fresh vegetable storage 143

6.8.   Storage structures for horticultural crops 144

6.9.   Transportation of horticultural crops 152

CHAPTER technological Processes for Fruits and Vegetables 157

7.   Introduction 157

7.1   Objectives of fruit and vegetable technological processes 157

7.2   Technological fruits and vegetable processes 158

7.3   Technological process operations for fruits and vegetables 172

7.3.1   Pre-processing procedures 172

7.3.2   Temporary storage of (fresh) fruit 179

7.4   Fruit and vegetable processing technologies 180

7.4.1   Drying and dehydration technology 180

7.4.2   Processing of fruit bars 181

7.4.3   Osmotic dehydration in fruit processing 182

7.4.4   Technology of semi-processed fruit products 185

7.4.5   Technology of "fruit in syrup" products 185

7.4.6   Technology for fruit juice 186

7.5   Recent trends in fruit processing 189

7.6   Introduction to food preservation 191

BIBLIOGRAPHY 192

Preface

The technological processes of harvesting, handling, processing, preservation and storage of horticultural crops cannot be fully appreciated without recourse to good understanding of the fundamentals of the biological nature of the crops, composition of the crop, crop utilization potentials as well as the nutritional qualities from the view point of their behaviour under prevailing or modeled atmospheric conditions.

This book is written to provide the students with a good understanding in fruits and vegetables handling, processing, and technological advances in preservation of fruits and vegetable from harvest till it gets to the consumer table or ended at the store shelf as finished products. Fruits and vegetables surfers the highest degree of deterioration at all levels of technological involvement right from maturity till shelving. This book is therefore packaged to advance knowledge and increase understanding of the nature of the fruits and vegetables in order to match up the principles and techniques of crops handling, processing and storage in order to minimize post harvest losses.

The book is in two parts:

Part 1: Concepts of fruit and vegetable handling

Part 2: Fruit and vegetable processing and storage methods

Finally, fruits and vegetable technologies, management options is intended to meet the needs of vocational personnel in market gardening, orchards management, floriculture, amenity horticulture, nurseries management and technology, landscape horticulture, and other areas including pomology, viticulture, oenology, including postharvest physiology

It is also designed to increase knowledge in students studying to acquire degrees and proficiency in certificate, diploma and higher levels of training in horticulture, agriculture and other related disciplines.

Bello, R. S.
480001, Nigeria

INTRODUCTION

Contents: Learning objectives and book outline

INTRODUCTION

This book provide students with the necessary theoretical and technological knowledge of fruits and vegetable processing and storage to minimize harvest and postharvest losses resulting from poor harvesting techniques and handling as well as postharvest management options. It will also provide students with practical skills for effective entrepreneurship in fruits and vegetable technologies conformable to international best practices.

Learning objective

This book is designed to provide the students with a good understanding of the principles and techniques of fruits and vegetables management technologies that minimizes post harvest losses and also to;

1. Know the basics of fruits and vegetables as horticultural products, properties and economic values
2. Understand the problems combating prospects of fresh products availability, technological development and transportation especially in tropics and within Nigeria context
3. Understand fruit and vegetables qualities and its determinants
4. Know the appropriate stage to harvest tropical fruits and vegetables.
5. Know the pre-harvest conditions that predispose fresh fruits and vegetables to post-harvest deterioration.
6. Understand tropical environment in relation to storage life, as well as methods of minimizing post-harvest losses in tropical fruits and vegetables
7. Understand simple techniques and equipments for preservation of fruit and vegetables.

Book outline

The book is written to effectively espouse knowledge in the following areas:

1. Factors that pre-dispose crops to post-harvest losses through weed invasion, pests attack as well as diseases.
2. Pre-harvest condition enhancing bruises and cracks in tropical crops such as tomato, bananas, plantain, pineapple, etc
3. Consequences of water stress to fruit development and maturity.
4. Influence of temperature on rate of respiration.
5. Influence of relative humidity on growth of fungi on stored fruit and vegetables.
6. Maturity and ripening indices for fruits and vegetables e.g.

   a. Change in taste of produce

b. Specific gravity of produce
c. Leaf senescence
d. Change in the colour of fruits.

7. Explain the advantages of tree-top ripening such as higher sugar content, lower acid content, etc.

    a. Explain the disadvantages of moisture, loss of weight, etc.
    b. Define the term mature green stage.
    c. Explain why fruits and vegetables are harvested at mature green stage e.g. When intended for long distance market.

8. Explain the factors that determine the quality in fruits e.g.

    a. Juice content
    b. Sugar content
    c. Acidity
    d. Flavour
    e. Nutrients and vitamins content
    f. Soluble solid content.

9. Explain the factors that determine quality in vegetables e.g.

    a. Fiber content
    b. Colour
    c. Freshness (water content) and turgidity
    d. Level of importance of effectively maintaining the factors to enhance produce marketability

10. Problems of vegetables and fruits in transit

    a. Identify fruits and vegetables in transit from northern Nigeria to southern Nigeria. E.g. Onions, tomatoes, peppers, carrots, cage, cauliflower, etc
    b. Identify fruits and vegetables in transit from southern Nigeria to northern Nigeria e.g. Oranges, bananas and plantain, telfaria (fluted pumpkins), okra, bitter leaf etc.
    c. Identify the different local and vegetables packaging materials e.g. open bags, baskets, crates etc.
    d. Identify the different transportation methods for fruits and vegetables e.g. Train, trailers, lorries, bicycles etc.

PART 1

CONCEPTS OF FRUITS & VEGETABLE HANDLING

FRUITS AND VEGETABLE TECHNOLOGIES

CHAPTER 1

PROPERTIES OF FRUITS & VEGETABLE

Contents: Introduction, fruits and vegetable differentials, compositions, economic values, and deficiencies

1.   Introduction

The horticultural products include fruits, vegetables, ornamental plants and flowers, plantation crops, aromatic and medicinal plants and spices. However, of particular interest to horticulturists and process engineers are the fruits and vegetables, its compositions, utilization, harvest and storage effects on spoilage and processing methods. This chapter takes a look at the properties of fruit and vegetables, its nutritional contents and utilization.

1.1 Fruits

Definitions

Fruit can be defined as 'the edible product of a plant or tree, consisting of seed and its envelope, especially the latter when it is juicy or pulpy'. The consumer definition of fruit would be 'plant products with aromatic flavours, which are either naturally sweet or normally sweetened before eating'.

Botanically, fruit are those portions of the plant which houses seeds. Therefore, such items as tomatoes, cucumber, eggplant, pepper and others would be classified as fruits. Those plants items that are commonly eaten with deserts are considered fruits. Fruits are seed-bearing structure of a flowering plant.

A fruit is actually a ripened ovary, a component of the flower's female reproductive structure. Fertilization of the egg within the ovary stimulates the ovary to ripen, or mature. Depending on the type of plant, the mature ovary may form a juicy, fleshy fruit, such as a peach, mango, apple, plum, or blueberry. Or it may develop into a dry fruit, such as wheat, corn, or rice.

1.2 Type fruits

Fruits are grouped into several divisions depending principally upon botanical structure, chemical compound and climatic requirements. Differences in flower structure result in several types of fleshy fruits .fruits can be classified by the reproductive setup of a flower or by natural occurrence.

Classification by reproductive setup

The reproductive setup of a flower classified fruits into the followings:

- *Simple fruits*: These fruits are produced by flowers containing one pistil, the main female reproductive organ of a flower.
- *Aggregate fruits*: These fruits are developed from flowers which have more than one pistil. They consist of mass of small drupes that develops from a separate ovary of a single flower. Typical example include pineapple
- *Multiple fruits*: These fruits are developed not from one single flower but by a cluster of flowers.
- *Accessory fruits*: These fruits are developed from plant parts other than the ovary.

Naturally occurring fruits

These are *fleshy or wet fruit* and *dry fruits*. Most of the fleshy fruits are edible and are eaten all over the world. Not all dry fruits are edible; nonetheless they are types of fruits. Some fleshy fruits underwent further processing and as such classified as *dried fruits*.

Characteristics and monitoring of these classes of fruits are as enumerated below.

1.2.1 Fleshy or wet fruits

The main fleshy fruits fund in nature is classified into the following groups:

1. Berries
2. Drupes
3. Pome
4. Hesperidiums or citrus fruits
5. Pepo (Melon)

6. Fleshy aggregate fruits
7. Fleshy multiple fruits

1. *Berries*

A berry fruit develops from an ovary containing one or more carpel. Each carpel contains one or more ovules, so berries typically contain more than one seed. These fruits have a soft epicarp and the mesocarp and endocarp is fleshly. There are names in the list of berries that actually are the true berries! While some of the names are not berries at all! The thin line that divides these fruits is the botanical definition which categorizes them as berries and the other simply being a common perception or understanding that underlines some fruits as berries even if they are far from the botanical definition to be categorized as such.

Botanical understanding of a berry: A berry is a fruit that has a fleshy, edible pericarp (fruit wall) produced from a single ovary that covers one or many seeds. In other words, it is a single fleshy ovary that grows into a juicy fruit and has no barrier in between the seed and the juicy part that is eaten.

Layman's understanding of a berry: In his understanding, any small, juicy, colourful, and fleshy fruit is referred to as berry.

True berries

A true berry has a relatively soft pericarp with a thin exocarp or skin. The following example of fruits conforms to the botanical definition of berries.

- *Tomato*: Tomatoes are the most common garden fruits and kitchen ingredient that are used in a number of culinary delights. Tomatoes come in several colours (red, orange, yellow, and green) and shape (round, plum, and cherry). Yellow and orange tomatoes are often lower in acid content than the more common red.

Figure 1-1: Tomato fruits

When a tomato is cut in half, it displays a distinct section, each segment representing a separate ovary compartment, or carpel, with many seeds.

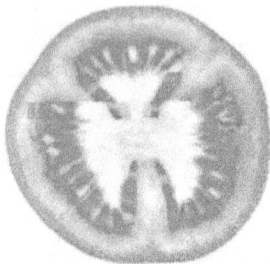

Figure 1-2: Cell arrangement in tomato fruits

- *Grapes:* Grapes are clusters small coloured fruits loaded with vitamin A, C, and B$_6$. They also contain potassium, calcium, magnesium, and folic acid.

Figure 1-3: Wine grapes

- *Pepper:* Pepper fruits are in different categories such as:
  - *Bell pepper:* Bell pepper is a cultivar group of the species *Capsicum annuum*. The cultivars of the plant produce fruits in different colours, including red, yellow, orange, green, chocolate/brown, vanilla/white, and purple. Bell peppers are sometimes grouped with less pungent pepper varieties as "sweet peppers".

Figure 1-4: Bell pepper

  - *Black pepper* (**Piper nigrum**) is a flowering vine in the family Piperaceae, cultivated for its fruit, which is usually dried and used as a spice and seasoning. The fruit, known as a peppercorn when dried, is approximately 5 millimetres (0.20 in) in diameter, dark red when fully mature, and, like all drupes, contains a single seed.

Figure 1-5: Black and white peppercorns

o *White pepper* consists of the seed of the pepper plant alone, with the darker coloured skin of the pepper fruit removed. This is usually accomplished by a process known as retting, where fully ripe red pepper berries are soaked in water for about a week, during which the flesh of the pepper softens and decomposes. Rubbing then removes what remains of the fruit, and the naked seed is dried.

Figure 1-6: Black, green, pink (*Schinus terebinthifolius*), and white peppercorns

o *Orange pepper or red pepper* usually consists of ripe red pepper drupes preserved in brine and vinegar. Ripe red peppercorns can also be dried using the same colour-preserving techniques used to produce green pepper.

• *Eggplants*: Eggplant is actually a fruit, specifically a berry, tear-shaped and usually purple-black vegetable. Although the most common type is large and dark purple, eggplant comes in many sizes (2-12 inches), shapes (oblong to round), and colours (white to green to purple). Its fleshy and substantive texture makes it a good replacement for meat. A member of the nightshade family, eggplant is related to the potato and tomato. Eggplant can be kept on the countertop for several days and widely used in cooking.

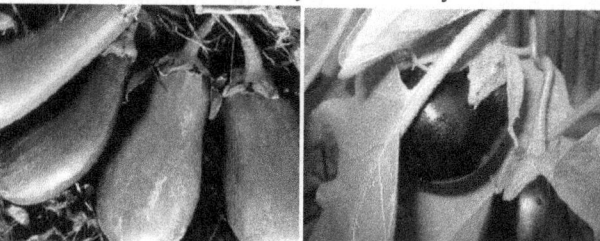

Figure 1-7: Eggplant

- *Okra*: Okra is a green, fuzzy, deeply ridged pod. Most common in warmer climates, it can also be grown in cooler areas if the proper variety and cultivation techniques are used. Okra is rich in Vitamins A, B6, C, K; thiamin; the foliate is rich in; calcium and fiber.

Figure 1-8: Okra fruit

Epigynous fruits (false berries)

These are developed from an inferior ovary as opposed to true berries that develop from a superior ovary. These are also called false berries. Typical example is blueberries.

- *Blueberries:* Blueberries are perennial flowering plants with indigo, dark-blue or purple - coloured berries. The fruit is 5–16 millimeters (0.20–0.63 in) in diameter with a flared crown at the end; they are pale greenish at first, then reddish-purple, and finally dark blue when ripe. They have a sweet taste when mature, with variable acidity.

Figure 1-9: Blueberries

They are used in jams, purée, juice, pies, and muffins. They contain high levels of antioxidants and can help prevent many diseases, like stomach ailments, heart deg

2. *Drupe*

Technically, these are not berries but taken to be. The difference being, they have a hard fruit wall and only one seed within. The fruit developed from an ovary with a single carpel and is characterized by an edible exocarp and mesocarp and an inedible, hard endocarp, or pit that encloses a single seed. They are also called stone fruits. The endocarp is hard and stony that fits closely around the seed. The mesocarp is fleshy and the fruit has thin, soft exocarp. The lists of fruits under the type of fruit are:

- *Cherry*: The cherry is the fruit of many plants of the genus *Prunus*, and is a fleshy drupe (stone fruit) consisting of a pulpy, globular drupe enclosing a one-seeded smooth stone.

Figure 1-10: Cherries

- *Avocados*: Avocado or alligator pear contains a single seed and have a green-skinned, fleshy body that may be pear-shaped, egg-shaped, or spherical. Commercially, it ripens after harvesting.

Figure 1-11: Avocado pear

- *Coconut*: Botanically coconuts are drupes, not a true nut with an edible endocarp and a very fibrous, inedible exocarp and mesocarp. Like other fruits, it has three layers: the exocarp, mesocarp, and. The exocarp and mesocarp make up the "husk" while the white fleshy part called coconut flesh makes up the endocarp. The mesocarp is composed of a fiber, called coir, which has many traditional and commercial uses.

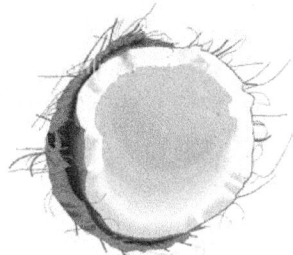

Figure 1-12: Drupes with hard exocarp; coconut

The hollow center contains a watery liquid called coconut juice which is often sipped straight from the coconut. The shell has three germination pores (stoma) or "eyes" that are clearly visible on its outside surface once the husk is removed. A full-sized coconut weighs about 1.44 kg (3.2 lb).

- *Plum*: Plums have a plump, round shape with a depression at the top where the stem attached. Plum skin is very smooth and shiny; it can have red, purple, or yellow colouration. Plum fruit is used to overcome constipation and has other health benefits.

Figure 1-13: Red (left) and yellow (right) plums

Other fruits in the group include peach, almond, apricot, and olive among others

3. *Pome*

Pome is a fleshy fruit composed of a mature ovary along with other enlarged parts of the flower. The fruit is developed from a compound inferior ovary. The ripened tissue around the ovary forms the fleshy edible part. The list of fruits under the types of pome is:

- *Apples:* In an apple the white, edible parts of the petals and sepals surround the ovary.

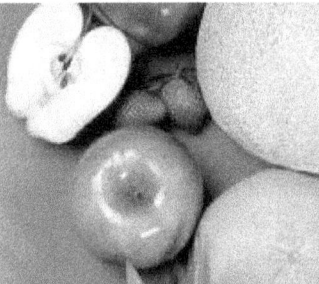

Figure 1-14: Apple fruits

Other fruits in the group include pears, quince, chokeberry etc.

4. *Hesperidiums or citrus fruits*

Citrus fruits are the ones that belong to the genus 'Citrus'. They are acidic, juicy and usually have a sharp taste. These fruits have thick, leathery exocarp and mesocarp. They have a juicy, pulpy endocarp arranged in a section of juice sacs from the ovary wall. All citrus fruits are rich sources of vitamin C which is required by the body to carry out different functions. Citrus fruits also have detoxifying properties, and hence, are used in some detoxification diet plans like the Master Cleanse (lemonade diet).

Figure 1-15: Citrus fruit

The list of fruit under the type of hesperidium includes:

- *Orange*: "Citrus aurantium" or orange, is the richest source of antioxidants of all fruits, stimulates the digestion, lowers the cholesterol and protects against cardiovascular disease

Figure 1-16: Orange fruit

- *Lemon:* The lemon tree bears large numbers of fruit throughout the year. The fruit is sold fresh, while the pulp of the fruit is used to produce frozen concentrate and fresh juice. The skin, or rind, contains oil used in the manufacture of perfume and lemon flavoring.

Figure 1-17: Lemon fruits

- *Grapefruit*: Grapefruit is a tropical citrus fruit, so named because it grows in grapelike clusters. Grapefruit is a cross between a sweet orange and a pummel

Figure 1-18: Grapefruits

Other fruits in the group include limes, citron, mandarin, celementine and tangerine

5. *Pepo (melon)*

There is a lot of debate on whether melon is a fruit or a vegetable. Scientifically, a melon is a fruit (based on the definition mentioned above). However, the confusion lies because of the fact that melons, cucumbers, squash, etc. belong to the same family, i.e., gourds. Melons are also classified into the broad category of fleshy fruits because they have an edible pulpy flesh. The berry has an outer wall or rind that is formed from receptacle tissue that is fused to the exocarp, the mesocarp and endocarp from the fleshy interior.

The lists of fruits under this category are:

- *Cucumber*: Cucumbers are long cylindrical green- smooth skinned fruit with edible seeds and soft white flesh. Smaller cucumbers (kirbys) are used for pickling. As they grow older, the seeds become bitterer and should be removed.

Figure 1-19: Cucumbers

There are three main varieties of cucumber: "slicing", "pickling", and "burpless".

Slicing cucumbers usually have skins that like long, round cylinders with a sweet flavour. Long, slender English cucumbers (also called Dutch or Japanese) have thin skins and contain few seeds.

Figure 1-20: Cucumber cut in longitudinal section

Pickling cucumbers are best used for making sweet or dill pickles.

- *Watermelon*: Watermelon fruit is very large, hard-skinned fruits with oval to round sweet, juicy smooth flesh. Watermelon has the highest water content. The skin can be solid green or green striped with yellow. The edible flesh is usually pink with many flat, oval, black seeds throughout. Seedless varieties also exist, as well as types with orange, yellow, or white flesh.

Figure 1-21 Watermelon

- *Cantaloupe*: The cantaloupes, also known as the rock melon, are berries with thick rinds. The cantaloupe has a thick, soft rind with a distinct netted appearance and a juicy, salmon-coloured pulp. Melons are large and have a tough outer cover.

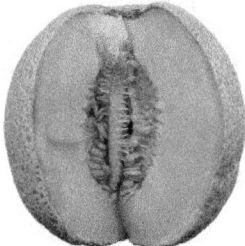

Figure 1-22: Rock melon

6. *Fleshy aggregate fruits*

Polydrupes: This is formed from the development of a number of simple carpels from a single flower. Few are dry fruits that are attached to a fleshy receptacle and the others are aggregation of simple fleshy fruits. The list of fruits under fleshy aggregate fruits includes:

- *Raspberry*: The raspberry is not a true berry but an aggregate fruit. The individual facets, or aggregate fruitlets, of a raspberry are called drupelets. In flowers with more than one pistil, the pistils are adjacent and the ovary of each pistil develops into a tiny fruit, or fruitlet.

Figure 1-23: Raspberry fruits

Raspberries and other brambleberries (blackberries, boysenberries, and mulberries) are soft fruits with tart, intensely flavoured juice. They come in a variety of colours red, black, purple, and gold. They grow on canes with or without thorns. Their high liquid content and thin membranes make them fragile. Raspberries are rich in Vitamins C and K; anthocyanins antioxidants (blackberries) and *fiber*.

- *Strawberry*: The fruit (which is not a botanical berry, but an aggregate accessory fruit) is widely appreciated for its characteristic aroma, bright red colour, juicy texture, and sweetness.

Figure 1-24: Strawberry fruits

7. Fleshy multiple fruits

Multiple fruits are called so because they are developed from multiple flowers. Every single flower produces a fruit, which is then assembled naturally as a whole. The individual ovaries from different flowers get clustered together forming a fruit. Therefore, what we see as a whole fruit is actually a bunch of fruits.

The list of fruits under the type of fleshy multiple fruits are:
- *Pineapple:* Pineapple fruits are compound oval fruits 6 to 8 inches long with spiky, robust leaves at the top of the fruit. The tough, waxy rind is green, brown, and yellow in colour

with a scale-like appearance. The flesh of the pineapple is juicy and yellow to white in colour.

Figure 1-25: Pineapple fruits

In pineapple, the ovaries of several flowers clustered together develop into individual fruitlets, which combine into a single larger fruit. The pineapple's meaty flesh and prickly skin form from the many flowers.

- *Breadfruit*: Breadfruit is a fruit though it is used more like a vegetable. It is a staple food in many areas such as the tropical countries. When cooked, it reportedly tastes just like potato. Many describe the taste to be similar to fresh-baked bread which hints at the choice of the name 'Breadfruit'.

Figure 1-26: Breadfruits

The flower develops into the compound fruit (or syncarp), oblong, cylindrical, ovoid, rounded or pear-shaped, 3½ to 18 in (9-45 cm) in length and 2 to 12 in (5-30 cm) in diameter.

In the green stage, the fruit is hard and the interior is white, starchy and somewhat fibrous. When fully ripe, the fruit is somewhat soft; the interior is cream coloured or yellow and pasty, also sweetly fragrant. The seeds are irregularly oval, rounded at one end, pointed at the other, about 3/4 in (2 cm) long, dull-brown with darker stripes.

Figure 1-27: Breadfruits

- *Jackfruit:* The flesh of jackfruit is starchy and fibrous, and is a source of dietary fiber. The flavour is comparable to a combination of apple, pineapple and banana.

Figure 1-28: Jackfruits

Other fruits in his group include; pomegranate, date etc.

Miscellaneous fruits

Miscellaneous fruits are can be categorized into either of the above-mentioned categories. They are native to only certain parts of the world and are not grown all over. Examples of fruits in these categories include:

- *Bananas:* Banana is a fleshy fruit with a soft epicarp; the mesocarp and endocarp is fleshly.

Figure 1-29: Banana

- *Currants:* These fruit are berries red, green, yellow, or black in colour. They are dried and used as raisins

Red currants: These are small round red or white berries that are used in making jams, tarts, and salads. They have high vitamin C, iron, potassium, and fiber.

Figure 1-30: Red currants

Black currant: This popular flavoured berry is like the red currant in appearance. It is used in jams, pies, ice creams, tarts, etc. Black Currant is packed with vitamin C. It also has high levels of potassium, phosphorous, iron, and vitamin B_5.

Figure 1-31: Black currants

1.2.2 Dry fruits

Dry fruits are often called nuts which usually have a hard dry covering called the shell that encloses an edible kernel.

Figure 1-32: Dry fruits

Dry fruits are classified by whether they remain intact at maturity, split open at one edge/seam when dry or open along the two seams to release seeds. Those fruits that remain intact at maturity are called the true nuts. Dry fruits that contain seeds in a seedpod that opens up and releases the seeds along one seam are known as *dehiscent fruits*. The indehiscent are those fruits that do not have a seed pot that opens. Let us have a look at different fruits that come under this group.

True nuts

The true nut such as chestnut is a dry fruit with one seed enclosed in a hard, thick pericarp. True nuts do not split open when ripe. Other examples of true nuts are the hazelnut, beechnut, and acorn.

hazelnut   chestnut   almond

Figure 1-33: True nuts

Dry dehiscent fruits

There are several types of dehiscent dry fruits such as

- *Follicle*: The fruit is developed from a single carpel ovary. It splits open from one side only. This type of fruit contains one or many seeds. The list of fruits under follicle includes: Columbine and milkweed.
- *Legume*: These fruits are dry dehiscent fruits such as peas and beans that have pods that split apart along the two edge seams, exposing the seeds that lie within. Some examples of fruits under legumes include:

Beans: Green beans (which may be other colours as well, and are also referred to as string beans) have long, edible pods and small inner beans. Green beans are rich in vitamins A, C, K; while the leaf or folate is rich in iron, potassium; lutein and fiber..

Figure 1-34: Green beans

Garden pea plant: Peas and other legumes are dry fruits characterized by one-chambered seed pods that split along two seams at maturity.

Sweet pea: Three main types are of sweet peas are available. *Shell peas* need to be removed from a pod before eating. *Sugarsnap peas* have sweet, edible pods and peas. *Snow peas* are an edible young flat pod with immature peas inside. Sweet beans are rich in Vitamins A, B6, C, K, thiamin, riboflavin, niacin, while the leaf or folate is rich in iron, potassium; lutein and fiber.

Figure 1-35: Sweet peas

- *Capsule*: The fruit develops from compound ovary with two or more carpels and the capsules dehisce. The fruits under this type of fruits are cotton, poppy and primrose.

Dry indehiscent fruits

Dry indehiscent fruits include:

- *Achene*: This is a small one-seeded fruit. The pericarp is easily separable from the seed coat. The fruits of buckwheat and sunflower family come under this type of fruit.
- *Samara*: These are one or two seeded achene-like fruits. They form wings from the outgrowth of ovary walls. The type of fruits under this group includes elms, ash and maple.
- *Caryopsis*: These are one-seeded small fruits that have pericarp completely fused to the seed coat. The type of fruits under caryopsis includes: Wheat, oats, rice,, corn, barley and rye
- *Nuts*: These are one-seeded dried fruits with a hard pericarp. The list of fruits under this category includes: Walnut, hazelnut, chestnut and, acorn

1.2.3 Dried fruits

Dried fruit refers to fruit from which the majority of the original water content has been removed *conventionally* or *forcefully* in the sun, wind tunnel dryers or any specialized drying equipment or dehydrators. Dried fruits retain most of the nutritional value of fresh fruits. The specific nutrient content of the different dried fruits reflects their fresh counterpart and the processing method. Examples of such fruits dried conventionally include raisins, dates, prunes (dried plums), figs, apricots, peaches, apples and pears. Many of such fruits as blueberries, cherries, strawberries and mango are infused with a sweetener (e.g. sucrose syrup) prior to drying to improve their eating values.

Figure 1-36: Date fruit (*Phoenix dactylifera* L.)

Advantages of dried fruits

Dried fruits have the advantage of being very easy to store and to distribute; they are readily incorporated into other foods and recipes, relatively low cost and present a healthy alternative to sugary snacks.

The greatest benefit of including dried fruits, with their unique combination of essential nutrients, fiber and bioactive compounds regularly in the diet is that it is a means to expand overall consumption of fruit and the critical nutrients they contain.

Importance of dried fruit

Dried fruit is important because of its sweet taste, nutritive value, and long shelf life. Today, dried fruit consumption is widespread. The following importance has been ascribed to regular consumption of dried fruits.

a. *Dietary value of dried fruit*
   i. Dried fruit and vegetables are vital to human diet and form an essential part of a balanced ration. Many species of mammals, birds, and insects rely on fruit as an essential component of their diet.
b. *Nutritional value*
   i. In general, all dried fruits provide essential nutrients and an array of health protective bioactive ingredients, making them valuable tools to both increase diet quality and help reduce the risk of chronic disease.
   ii. Individuals who regularly eat generous amounts of these foods have lower rates of obesity, cardiovascular disease; sever cancers, diabetes and other chronic diseases.
   iii. Dried fruits are important sources of vitamins, minerals and fiber in the diet

iv. Dried fruits retain most of the nutritional value of fresh fruits, and so are included with fresh fruit in dietary recommendations.

v. Dried fruits provide essential nutrients making them valuable tools to increase diet quality and help reduce the risk of chronic disease.

c. *Health value*

i. Dried fruits provide a wide array of bioactive components or phytochemicals that play a role in health and longevity and have been linked to a reduction in the risk of major chronic disease.

ii. Dried fruits are an important source of antioxidants in the diet that can lower oxidative stress and so prevent oxidative damage to critical cellular components..

iii. Dried fruits such as dried plums provide pectin, a soluble fiber that may lower blood cholesterol levels.

iv. Dried fruits such as raisins are a source of prebiotic compounds in the diet that contribute to colon health.

v. Dried fruits contain organic acids such as tartaric acid (raisins) and sorbitol (dried plums) that help maintain a healthy digestive system.

vi. Traditional dried fruit have a low to moderate Glycemic Index (GI) – a measure of how a food affects blood sugar levels. GI measures an individual's response to eating a carbohydrate-containing food (usually 50 grams of available carbohydrates) compared to the individual's response to the same amount of carbohydrates from either white bread or glucose.

vii. Dried fruit promote healthy teeth and gums contrary to longstanding popular perception that dried fruits such as raisins promote cavities, recent studies indicate that they may benefit oral health.

viii. Dried fruits, particularly dried plums, promote bone health. Research conducted with dried plums indicates that they have a role in supporting bone health.

ix. Dried fruits promote digestive health. There is considerable research supporting the role of dried fruit in regulating bowel function and maintaining a healthy digestive system.

1.3   Tropical fruits

These are assorted native fruits that naturally grow in the wild, domesticated or cultivated and found widely spread/distributed in the tropical geographic zone of the world climate. Examples of such fruits include but not limited to orange, pineapple, lime, mango avocado etc.

Figure 1-37: Tropical fruits

1.4   Vegetables

Certain foods commonly termed vegetables, are herbaceous plant usually cultivated for food including tomatoes, squash, peppers, and eggplant, technically are fruits because they develop from the ovary of a flower.

Figure 1-38: Typical vegetables

Definitions

Botanically many crops, defined as vegetables, are fruits (tomato, melons, squash, peppers, and eggplant etc.) technically are fruits because they develop from the ovary of a flower. Morphologically and physiologically the fruits and vegetables are highly variable, may come from a root, stem, leaf, immature or fully mature and ripe fruits. They have variable shelf life and require different suitable conditions during marketing. The classification of fruits and vegetables is arbitrary and according to usage.

Classification of vegetable

Vegetables are generally classified according to the source of the edible plant part(s). Examples of these groups include:

1.   *Root vegetables*

Examples of root vegetables include carrots, beetroot, radish etc.

- *Carrot*: Carrots are crisp, firm, and small to medium root vegetable grown for its long tapered to small round ball-roots. Some come in orange (by far the most common), purple and red varieties with varying degree of nutritional values. For instance the dark orange carrots have more vitamins A.

Figure 1-39: Root vegetable (Carrots)

2. *Leafy green vegetables*

Leafy green vegetables are the commonest food you can eat regularly to help improve your health because leafy vegetables are brimming with fiber along with vitamins, minerals, and plant-based substances that may help protect you from heart disease, diabetes, and perhaps even cancer. Examples include silver beet, lettuce, spinach, cabbage

- *Lettuce*: **Lettuce** is an annual plant of the aster or sunflower family Asteraceae. It is most often grown as a leaf vegetable, but sometimes for its stem and seeds. Lettuce is most often used for salads, although it is also seen in other kinds of food, such as soups, sandwiches and wraps; it can also be grilled.

Figure 1-40: Lettuce

- *Cabbage*: Cabbage's tightly layered leaves form a compact head surrounded by darker outer leaves. Common varieties are green, red, crinkly savoy, Chinese, bok choy, and napa. Cabbages are generally interchangeable in recipes, Chinese cabbages cook in less time.

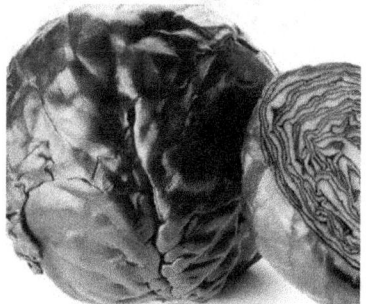

Figure 1-41: Cabbage

3. *Stalk vegetables*

Stalk vegetables are those vegetable with rigid, compact, straight stalks with high luster grown in for fresh market sales. Examples include (Celery chard, fennel and asparagus);

- *Celery:* a plant variety in the family Apiaceae, commonly used as a vegetable. The plant grows to 1 m (3.3 ft) tall. The leaves are pinnate to bipinnate with rhombic leaflets 3–6 cm long and 2–4 cm broad. The flowers are creamy-white, 2–3 mm in diameter, and are produced in dense compound umbels. The seeds are broad ovoid to globose, 1.5–2 mm long and wide.

Figure 1-42: Celery

- *Asparagus:* A perennial, asparagus are spear-like shoots that come in three main varieties: green, the most common; white, for which the green variety is field-blanched; and purple, an extra sweet and tender variety that turns green when cooked. **Nutrients:** Vitamins A, C, **K**, thiamin, riboflavin, folate; iron; fiber.

Figure 1-43: Asparagus

4. *Bulb vegetables*

These are vegetables grown mainly for their bulbs, the underground structure where the plant's nutrient reserves are stored. The main edible part of these vegetables is their bulb. Common examples include garlic, onion, chives and leek.

- *Onion:* Varieties of this bulb vegetable include yellow, red (milder and sweeter), white, pearl, Spanish (very mild), and sweet onions.

Figure 1-44: Onion bulbs

White onion: this onion is widely used as a flavoring ingredient,; it is often eaten raw or deep-fried in rings.

Red onion: This is the sweetest of the onions; it is often eaten raw, in salads or sandwiches.

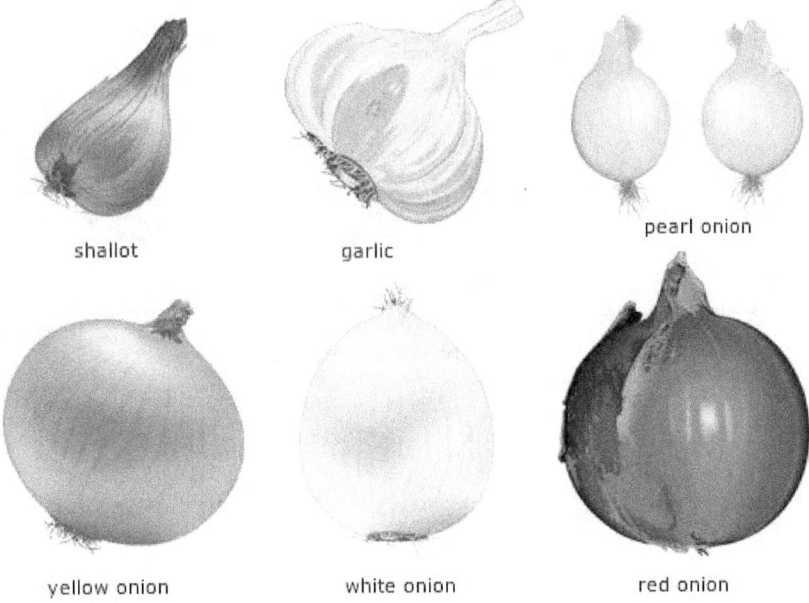

shallot   garlic   pearl onion

yellow onion   white onion   red onion

Figure 1-45: Onion bulbs

Pearl onion: These are small white onion picked before fully ripe; it is primarily used to make pickles or as an ingredient in stews.

Garlic: The bulb of garlic is composed of bulblets called cloves; the germ at its center can make garlic difficult to digest.

Yellow onion: The most common of all the varieties of onion widely used as a flavoring ingredient, either raw or cooked. It is also the essential ingredient in preparing onion soup.

Shallot: It has a more subtle flavor than the onion or the chive; it is eaten raw or cooked and often used as a flavoring ingredient in sauces.

5. *Tuber vegetables*

These are root tubers eaten like vegetables; they consist of underground growths containing the plant's nutrient reserves. Examples of tuber vegetable include taro, yam and cassava.

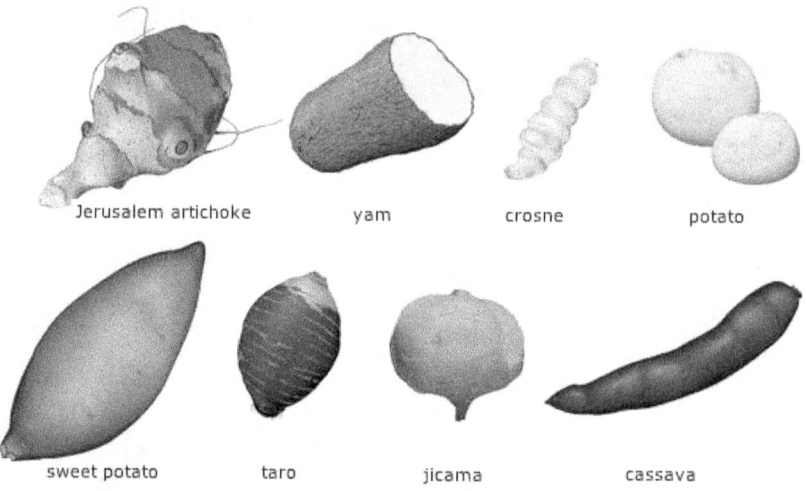

Figure 1-46: Tuber vegetables

6. *The "inflorescent vegetables*

Examples of inflorescence vegetable include artichokes, cauliflower, broccoli rabe and broccoli.)

Figure 1-47: Inflorescence vegetables

7. *The fruit vegetables*

These are not actually vegetables, but botanical definition of fruit, which is, "the ripened ovary of a plant and its contents... and seeds together with any structure with which they are combined" has classified them as fruits. These are olives, squash, avocados, cucumbers, peppers, tomatoes, okra, and eggplant.

Figure 1-48: Fruit vegetables

1.5   Importance of fruits and vegetables

The following importance is derivable from fruit and vegetables

Dispersal & propagation: Fruits plays a critical role in dispersing seeds, increasing the likelihood that at least some will land in an environment favourable for germination, or sprouting, which helps to perpetuate the plant species.

Medicinal value: Fruits are essential in the diet to prevent certain diseases. For example:

a. Scurvy, a potentially fatal disease marked by swollen joints, inflamed gums, and weakness, resulting from lack of vitamin C, the vitamin found in particularly high concentrations in oranges, lemons, and limes.
b. Unprocessed grains and legumes, along with other foods supply thiamine (vitamin B_1), which prevents beriberi, a potentially fatal disease of the nervous system.
c. Many fruits are also rich in vitamin A, which prevents night blindness, supports the immune system, helps bones grow, keeps skin healthy, and plays many other indispensable roles in maintaining health.

1.6   Nutritional content of fruits and vegetables

Vegetables are high in mineral substances including K, Na, Ca, Mg, Fe, Mn, Al, P, Cl, and S. mineral contents of particular importance are K, Fe, and Ca.

Vegetables such as green beans, cabbage, onions, and beans are high Ca. Spinach, carrots, cabbage, and tomatoes are particularly rich in minerals.

Vitamin A is provided through beta-carotene found in orange and yellow vegetables and green leafy vegetables. Vitamin B also is found in fruits and vegetables in significant quantities.

Fruits rich in minerals include strawberries, cherries, peaches and raspberries. Citrus papaya, mango, kiwifruit, tomatoes, cabbage and green peppers are high in vitamin C (ascorbic acid). Apples are high in Fe.

1.7   Comparison between fruit and vegetable

Similarities

Fruits and vegetables have many similarities with respect to their compositions, method of cultivation and harvesting, storage property and processing. Some generalization can be made comparing fruits with vegetables as follows:

1. Most fresh vegetables and fruits are high in water content, low in protein and fat. In these cases, water content will generally be greater than 70% and frequently greater than 85%.
2. Fruits and vegetables are important sources of both digestive and indigestive carbohydrates. The digestive carbohydrates are present largely in the form of sugar and starch, and indigestible cellulose provides roughages which are important to normal digestion.
3. Fruits are important sources of minerals and certain vitamins especially A and C. Citrus fruits for example are excellent sources of vitamins C, they contain natural acids, such as citric acid in oranges and lemons, maltic acid of apples and tartaric acid of grapes. These acids give the fruits its tartness which slows down bacterial spoilage.
4. Fruits and vegetables are good sources of enzymes which slows down or promotes most bacterial reactions occurring in the cells, thus they exhibits some of these properties:

    a. Control of ripening in fruits. In many cases, to the point of storage such as in soft melon and over ripped banana.
    b. They may be responsible for changes in flavour, colour, texture and nutritional properties.

5. Heating process in fruits and vegetable processing are designed not only to destroy micro-organisms but also to deactivate enzymes and so improve the fruits' storage stability. Note that Enzymes have an optimal temperature around $+50^\circ c$ where their activities are at maximum. Heating beyond this optimal deactivates the enzymes.

Table 1-1: Differences between fruits and vegetables

	Vegetables	Fruits
1.	Vegetable generally means the edible parts of plants	Fruits are the ripened ovaries of flowering plants.
2.	Vegetables widely classified as foods that are eaten as part of a meal's main course	Fruits are generally categorized as foods that are eaten for dessert or as a snack.
3.	Most vegetables are less sweet because they have much less fructose	Most fruits are sweet because they contain a simple sugar called fructose

1.8   Fruit development and growth regulator

Living organisms are complex structural and biochemical entities. Why a particular species differs from others and how a complex organism arises from a single cell is not clearly understood. It is clear that a high degree of organization of growth processes must exist to account for the orderly development of plants.

A few organic compounds that influence certain biochemical and physiological reactions have been isolated from the wide array of organic and inorganic compounds found in plants. A growing body of evidence suggests such compounds are of general occurrence in plants, active at exceedingly low concentrations, and under genetic control. These compounds, which act as chemical directors of growth processes, are known as plant-growth substances or plant-growth regulators.

Figure 1-49: Trademark growth regulators

Among several growth regulator employed in plant treatment are; indole-3-acetic acid (IAA), gibberellic acid (GA3), aminoethoxyvinylglycine (AVG), s-abscisic acid (s-ABA), 6-benzyladenine (6-BA), cytokinin forchlorfenuron (CPPU) and coconut milk.

1.8.1   Relevant engineering aspects of growth regulators

Fruit abscission: Abscission causes serious fruit loss, especially during the latter part of the harvest season. Citrus fruits have no well-defined physiological maturity stage, and as such, it affects its harvest. Harvest of oranges and grapefruit begins when the fruits are mature enough to be eaten and usually ends four to six months later. Growth regulator has the potential to improve abscission for an appreciable length of time for a greater percentage of the fruit to be harvest at once.

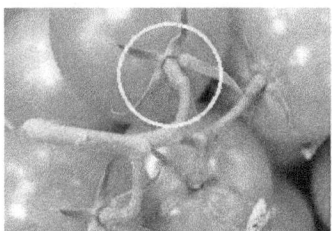

Figure 1-50: Point of abscission in tomato fruit

Fruit size: Growth regulator also affects the size of fruits produced (positively or negatively). For example, when small citrus fruits are treated with high concentrations of 2, 4-D, many of them becomes misshapen (irregular shape). Treated navel oranges may develop thick rinds and protruding navels or become cylindrical in shape.

Figure 1-51: Uniform size of jack fruit

Small seed-like structures, mainly enlarged seed coats, develop in navel oranges if high concentrations are applied to small fruits. For example, MaxCel growth regulator stimulates cell division, resulting in increased fruit size at harvest.

Figure 1-52: Fruit size growth regulator trademark

Improving fruit set: Fruit set implies a tree producing relatively same sizes of fruit at maturity. It is a term used in botany and horticulture to describe transition of flower to *fruit* with or without stimulus of pollination, leading to swelling and growth of ovary. A means of obtaining improved fruit set in fruit would prove of enormous commercial benefit, and plant growth regulators could provide the means of bringing about such responses.

 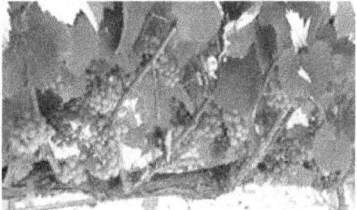

Figure 1-53: Fruit sets

Prestige® is a plant growth regulator product containing the cytokinin forchlorfenuron (CPPU) registered for table grapes in Australia. Prestige increases fruit set and size by stimulating cell division and improves fruit firmness, uniformity, cap stem attachment, and harvest flexibility.

Figure 1-54: Fruit set growth regulator trademark

Maturity: Plant-growth regulators are a potential means of regulating various aspects of fruit maturity. The development of methods for bringing about early fruit maturity or delaying fruit maturation could prove of considerable advantage for numerous citrus varieties grown under a broad range of climatic conditions and subject to various marketing demands.

Figure 1-55: Fruit dropping at maturity

Improved product quality: ProGibb® is a gibberellic acid (GA3) plant growth regulator product used in a wide variety of crops to improve quality and value. ProGibb is a highly effective growth promoter that increases size and quality of fruits, vegetables, and other crops. ProGibb can also play a role in the regulation of other plant processes such as flowering, seed germination, dormancy, and senescence.

Figure 1-56: Product quality improver

Fruit storage: The keeping quality of fruit is a primary concern of the citrus industry. In general, the maximum interval between harvest and consumption of oranges and grapefruit is two months. Lemons may be kept in lengthy storage before being placed in cartons for shipment. Length of storage of lemons ranges from a few weeks for yellow-coloured fruit to five or six months for dark-green fruit. Extensive research is being conducted on means of prolonging storage and reducing storage disorders.

Figure 1-57: Fruit storage

Flowering: Among the major environmental factors known to be critical to flowering of most plants are temperature, soil moisture, and photoperiod. Since qualitative and quantitative changes of plant-growth substances occur during flower induction and development stages, an engineering approach to these factors has helped improve their development.

Figure 1-58: Tomato fruit development from flower stage

For example, PinCor® is a plant growth regulator containing aminoethoxyvinylglycine (AVG) that inhibits the biosynthesis of ethylene in plant tissues and flower induction in pineapple.

Vegetative effects: Some information is available on the influence of plant-growth substances on the vegetative growth of citrus. One vegetative growth response, the use of maleic hydrazide (MH) to enhance dormancy and reduce susceptibility to frost, is already being used commercially to some extent in Florida. Plant-growth substances have played a major role in enhancing the rooting of cuttings from a wide range of plants.

Figure 1-59: Tomato fruit development from flower stage

Promalin® is a mixture of two naturally occurring plant growth regulators: gibberellic acid 4 and 7 (GA4+7), which cause cell enlargement and elongation, and 6-benzyladenine (6-BA), which promotes cell division. **Release®** activates the germination process and stimulates active seedling development in the early stages of seedling growth, improving germination rate, accelerating plant emergence, and establishing more uniform stands.

Residues of plant-growth regulators

The possibility of residues of plant-growth regulators remaining in citrus fruit at harvest presents a problem that will become increasingly important as commercial applications are developed for these compounds.

1.9   Fruit market

A visit to any local farm market where you have access to super fresh fruits and vegetables, juices, eggs, fish, and meat is a wholesome experience. When the farmers themselves bring their fruits and vegetables to market, you know they are fresh. No cold storage, ripening with ozone, oddly perfectly sized or shiny produce; just the regular old odd shapes and sizes, ones that smells and look good.

Figure 1-60: Local fresh fruits market

By-passing the chain of suppliers that bring your goods from the farm to the supermarket prevents additional costs being added to your food. When produce is produced locally in season, it automatically costs less to produce.

At the market you find foods that are in season, not food that's been sitting around in a chilled warehouse in storage. Eating foods in season is when it is most flavourful, nutritious and most affordable. When food is not in season, it is either grown in a glasshouse or imported from other parts of the world. Both affect the taste.

Farmers markets are great opportunities for small suppliers to rent a space, unpack a table, and launch their unique goods. They can easily afford a stall without the additional costs of selling through a shop. This means they can bring their specialized products to market and give us the gift of extending our standard purchases.

CHAPTER 2

FRUITS & VEGETABLE HARVEST TECHNOLOGIES

Content: Introduction , changes in fruit and vegetables, maturity, fruits and vegetable handling and harvest technologies

2   Introduction

Harvesting of fruits and vegetables at proper stage of maturity is of paramount importance for attaining desirable quality. The level of maturity actually helps in selection of storage methods, estimation of shelf life, selection of processing operations for value addition etc. Fruits harvested too early may lack flavour and may not ripen properly, while produce harvested too late may be fibrous or have very limited market life. Similarly, vegetables are harvested over a wide range of physiological stages, depending upon which part of the plant is used as food. For example, small or immature vegetables possess better texture and quality than mature or over-mature vegetables.

2.1   Factors affecting changes in fruits and vegetable

Composition of fruit changes with the degree of maturity prior to harvest and the condition of ripeness, which is progressive after harvest and is further influenced by storage conditions. Fresh vegetable are living organisms and there is a combination of life processes after harvest. Feasible changes that occur in fruits and vegetable which affects their ultimate qualities are largely predetermined at harvest.

Such factors that will affect fruit at harvest include; choice of cultivar, rootstock, weed and pest controls, irrigation practice, pruning systems and types of manpower training. Some of these factors are briefly explained below.

Choice of cultivar

Certain cultivar, when ripening could develop postharvest complications resulting from microorganisms, pests and diseases. For instance, it has been claimed that avocado tree has a major influence on the amount of post harvest rots that avocado fruit develop when ripe (Whiley et al., 2007).

Rootstock

Rootstock is one of the biggest importances among biotic factors influencing fruit quality. It affects fruit size, flesh firmness, soluble solids concentration and storage potential. However, understanding the role of the rootstock is not an easy task, because of its interactive nature (Castle 1995).Fruit quality characteristics vary according to the rootstock type. Rootstock has been associated with plant physiological composition. The differences in physiology lead to difference in fruit mineral profiles, changes in starch accumulation profile (Lahav and Whiley, 2001; Whiley et al., 1997) and leaf anti-fungal diene levels (Willingham et al., 2001). These changes lead to greater or lesser crops

Weed control practices

Weed and pest control practices is one of the factors that could adversely affect postharvest quality. Among the weed control practices in use, cultural/mechanical and chemical control practices are the most prevalent.

Cultural weed control: This is the use of physical or mechanical means to root out weed or plant totally from the field. This includes such practices as tillage, hoeing, hand picking, digging, mowing, burning, flooding and mulching. Cultivator is used for mechanical weed control.

Figure 2-1: A range of manual weeders

Chemical weed control: The use of chemicals for the control of weeds offers the greatest possibilities for weed control. Chemical weed control functions on the basis that certain chemicals are capable of killing weed without significant injury to other plants (crops). Such chemicals could be selective or non-selective. Insecticides are applied to soil and crops in the form of granules, dust, or liquid sprays.

Figure 2-2: Boom sprayers for agricultural crops

A variety of mechanical spraying and dusting equipment is used to spread chemicals on crops and fields; the machinery may be self-powered, or drawn and powered by a tractor. In areas where large crops of vegetables and grain are grown, airplanes are sometimes used to dust or spray pesticides.

Irrigation practice

Increasing demands on the production of fruits and vegetable round the year and the introduction of various fruits and vegetable into new growing regions had necessitated the increased focus on irrigation in gardening and orchard farming. However, the type of irrigation practice and quality of water used in irrigation do affect the product quality at harvest. An understanding of plant water relations and soil water management practices is essential to effective irrigation use to produce consistent yields of high-quality products. Effective irrigation management results in better control of plant growth and more efficient and economical crop production.

Pruning systems

Pruning, thinning, spraying are typical examples of fruit tree management methods and control systems. For instance, *the size and height of fruit determines the type of tree control system which are determinants in the productivity level of plant.*Pruning is a valuable method of controlling tree growth, improving and increasing flowering rate, fruit color, soluble solids concentration (SSC), flower bud formation and decreasing titratable acid content (TA) (İkinci, 199; Tehrani and Leuty, 1987; Mizutani et al, 2000).

Types of manpower training

Manpower training is largely concerned with labour supply in handling fruit and vegetables from growing stages to harvest. Thus it is important to know how many people are coming onto the labour field, their educational status, training received, levels; their age etc. manpower requirement is a concern in determining what training needs there are so that the labour supply can be shaped to meet the demands of the farm.

2.2   Maturity in fruits and vegetable

Maturity in fruits and vegetable is defined as a stage in their development when all the physiological processes have fully developed to its peak. At this stage, the fibrous and the non fibrous materials constituting the fruit has reached its peak. When harvest is done at this stage, the fruit can be processed into desired product or kept for distance transport.

The word mature is derived from Latin word '*Maturus*' which means ripen. The maturity stage at harvest is paramount in determining the final product quality after postharvest ripening. It is that stage of fruit development, which ensures attainment of maximum edible quality at the completion of ripening process.

Maturity at harvest is the most important factor that determines storage-life and final fruit quality. Immature fruits are more subject to shriveling and mechanical damage, and are of inferior flavour quality when ripe. Overripe fruits are likely to become soft and mealy with insipid flavour soon after harvest. Fruits picked either too early or too late in their season are more susceptible to postharvest physiological disorders than fruits picked at the proper maturity.

Maturity level

The level of maturity actually helps in selection of storage methods, estimation of shelf life, selection of processing operations for value addition etc. Maturity has been divided into three categories i.e. physiological maturity horticultural maturity and commercial maturity.

1. *Physiological maturity*: Physiological maturity refers to that stage of development when maximum growth has occurred and proper completion of subsequent ripening can occur even if the product has been harvested. This is the stage when a fruit is capable of further development or ripening when it is harvested i.e. ready for eating or processing. For instance, physiological maturity in bell pepper is reached when seeds become hard and the internal cavity of fruit starts colouring (Figure 2-3).

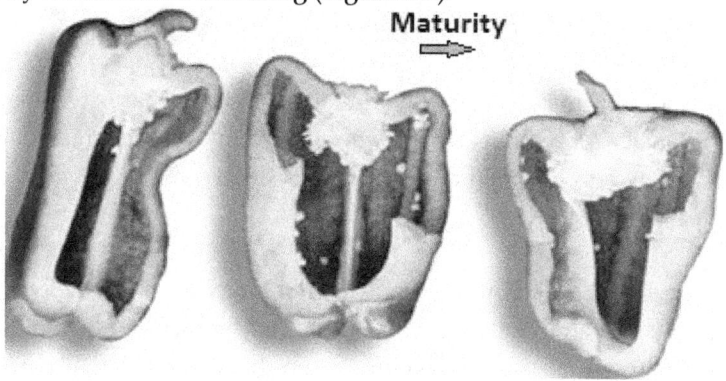

Figure 2-3: Physiological maturity in bell pepper

2. *Horticultural maturity*: This refers to the stage of development when plant and plant part possesses the pre-requisites qualities for use by consumers for a particular purpose i.e. ready for harvest.

3. *Commercial maturity* is that stage of development of a fruit or vegetable that is required by the market (retailer or consumer); it may have little relation to physiological maturity and may occur at various stages of ripeness depending on individual consumer preference.

Over maturity or over ripening is the stage that follows commercial maturity and is then the fruit softens and loses part of its characteristic taste and flavour. However, this is the ideal condition for preparing jams or sauces.

2.2.1 Maturity index and measurement

Not all fruit in an orchard, or even on the same tree, will ripen and turn colour at the exact same time, so a maturity index was developed to help producers numerically categorize their fruit's maturity level. A maturity index number allows producers to evaluate their varieties under their own specific growing conditions over a number of years.

Importance of maturity indices

Maturity index is useful in achieving one of the following objectives

- Ensure sensory quality (flavour, colour, aroma, texture) and nutritional quality.
- Ensure an adequate postharvest shelf life.
- Facilitate scheduling of harvest and packing operations.
- Facilitate marketing over the phone or through internet.

Determination of maturity index

Maturity index can be evaluated by one of the following methods:

1. *Determination of maturity index by colour changes*

The loss of green colour of many fruits is a valuable guide to maturity. There is initially a gradual loss in intensity of colour from deep green to lighter green and with many commodities, a complete loss of green colour with the development of yellow, red or purple pigments.

Ground colour as measured by colour charts, is useful index of maturity for apple, pear and stone fruits, but is not entirely reliable as it is influenced by factors other than maturity. For some fruits, as they mature on the tree, development of blush colour, that is additional colour superimposed on the ground colour, can be a good indicator of maturity. Examples are red or red-streaked apple cultivars and red blush on some cultivars of peach.

Objective measurement of colour is possible using a variety of reflectance or light transmittance spectrophotometer. Colour perception depends on the type and intensity of light, chemical and physical characteristics of the commodity, and person's ability to characterize colour. Although human eye is used to evaluate colour, however, results can vary considerably due to human differences in colour perception.

Therefore, an instrument (objective method) is used to provide a specific colour value based on the amount of light reflected off the commodity surface or light transmitted through the commodity. Maturity Index (MI) is determined as follows:

1. Take a random sample of about 2kg of fruit from several trees (high and low) of the same variety in the area where harvest is imminent. Collect all the fruit from a branch here and there rather than individual fruits.
2. Randomly select 100 fruit out of the sample bucket. Repeat 3-10 times until all the fruit is gone.
3. Separate each 100 fruit sample into eight colour categories based on the colour chart below:
   - 0 = Skin colour is deep green-fruit is still hard
   - 1 = Skin colour is yellow-green -fruit starting to soften
   - 2 = Skin with less than half the fruit surface turning red, purple, or black
   - 3 = Skin colour with greater than half the surface turning red, purple, or black
   - 4 = Skin colour all purple or black with all white or green flesh
   - 5 = Skin colour all purple or black with less than half the flesh turning purple
   - 6 = Skin colour all purple or black with greater than half the flesh turning purple
   - 7 = Skin colour all purple or black with all the flesh purple to the pit

4. Multiply the number of fruits in each colour category by the number of that colour category (0 to 7)
5. Add all the numbers together and divide by 100:

$$MI = \frac{Ax0 + Bx1 + Cx2 + Dx3 + Ex4 + Fx5 + Gx6 + Hx7}{100} \quad ....2.1$$

Letters (A-H) represents number of fruit in each category.

For example, the separation of the fruits in tray on the left of Figure 2-4 is shown in the trays to the right. The fruits were separated as follows

Figure 2-4: Determining maturity index by shape of shoulder

A: 5 green x 0 = 0 ,

B: 20 yellow − green x 1 = 20 ,

C: 20 $<\frac{1}{2}$ colour turning red, purple or black x 2 = 40 ,

D: 28 $<\frac{1}{2}$ colour turning red, purple or black x 3 = 84 ,

E: 12 black/white flesh x 4 = 48 ,

F: 8 black $>\frac{1}{2}$ purple flesh x 5 = 30 ,

G: 5 Black $>\frac{1}{2}$ purple flesh x 6 = 30 ,

H: 2 black flesh to pit x 7 = 14 ,

$$MI = 0 + 20 + 40 + 84 + 48 + 30 + 14 = 176 = \frac{276}{100} = 2.76$$

2. *Determination of maturity index by size and shape*

Maturity of fruits can be assessed by their final shape and size at the time of harvest. Fruit shape may be used in some instances to decide maturity. For example, the fullness of cheeks adjacent to pedicel may be used as a guide to maturity of mango and some stone fruits (Figure 2-5).

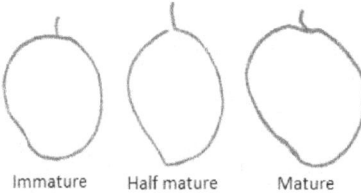

Immature   Half mature   Mature

Figure 2-5: Judging mango harvest maturity by shape of shoulder

Although there are very good predictive models for fruit size based on early season temperatures and fruit size at an early growing stage, these are not good indicators of maturity. Small fruit about 60 days from full bloom will usually remain small at harvest whether approaching maturity or not.

3. *Determination of maturity index by firmness*

As fruit mature and ripen they soften by dissolution of the middle lamella of the cell walls. The degree of firmness can be estimated subjectively by finger or thumb pressure, but more precise objective measurement is possible with pressure tester or penetrometer. Fruit firmness is probably the most reliable indicator of maturity in pears. There is a firmness pressure range classified to determine fruit maturity. Those fruits destined for immediate shipment or short-term cold storage is typically harvested at the lower end of the range. For longer control atmospheric storage of fruit or late marketing, harvest when fruit firmness is at the upper end of the pressure range.

4. *Determination of maturity index by soluble solids*

Amount of soluble solids has been an old standby for fruit maturity for many years, however; soluble solids are not a good maturity predictor due to variability between years, orchards, and even within the tree. The percent of soluble solids does play a role in stored fruit quality.

5. *Determination of maturity index by acidity level*

Acid levels in fruit affect taste and flavour but are a poor prediction of fruit maturity. Acid levels change yearly and usually also varies between orchards. Higher acid levels do positively affect taste and therefore, should improve quality and marketability.

2.3   Ripeness

The word ripe was derived from Saxon word '*Ripi*', which means gather or reap. This is the condition of maximum edible quality attained by the fruit following harvest. Only fruit which becomes mature before harvest can become ripe. When a product has attained the state or stage of maturity, it is said to be mature or ripe.

This is not a physiological process in fruit development but a declining stage in the life of the fruit. Ripening is thus the change of the green chlorophyll content of the fruit into various colourations as an indication of quality degeneration. Thus ripening is a state of degeneration of fruit through changes in colour, texture, firmness, sugar content, acidity content etc. for instance, the degree of ripening in tomato (from left to right of Figure 2-) is described as follows: 1) Mature green; 2) Breaker; 3) Turning; 4) Pink; 5) Light red and 6) Red. Due to its climacteric ripening characteristics, the tomato fruit reaches stage 6 even when harvested at maturity stage 1.

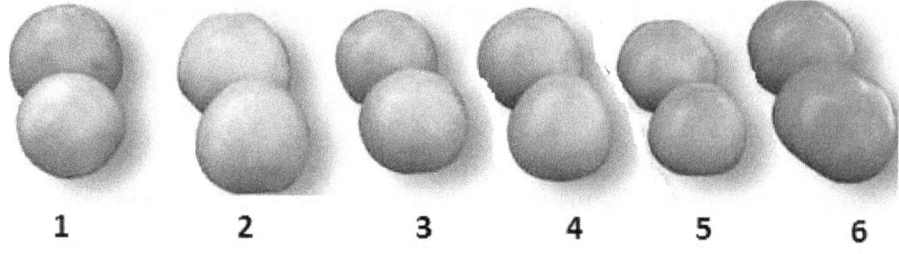

| 1 | 2 | 3 | 4 | 5 | 6 |

Figure 2-6: Degree of ripening in tomato

Note that freshly harvested horticultural products remain alive after harvest, in contrast with other food products such as cereals, meat, and milk-based items. After harvest, horticultural products do not remain in a constant condition but continue to develop through the following processes that are genetically predetermined:

$$Maturation \rightarrow Ripening \rightarrow Senescence \rightarrow Death$$

Each of the processes is explained as follows:

Maturation

This is the developmental process by which the fruit attains maturity. It is the transient phase of development from near completion of physical growth to attainment of physiological maturity. There are different stages of maturation e.g. immature, mature, optimally mature, over mature.

Ripening

Final eating quality is critically dependent on harvesting at the correct maturity stage, so that normal ripening can occur with the concomitant development of flavour, texture, aroma and juiciness required by consumers. In many fruits, ripening occurs either on or off the tree. Optimum eating quality for many vegetable crops is attained before full maturity. Examples include peas, green beans, sweet corn, asparagus, and leafy vegetables; if these products are left attached to the parent plant and not harvested at the correct time, their quality is much reduced.

Ripening involves a series of changes occurring during early stages of senescence of fruits in which structure and composition of unripe fruit is so altered that it becomes acceptable to eat. The developmental processes of *maturation* (maturity) and *ripeness* merge and overlap.

Ripening is a complex physiological process resulting in softening, colouring, sweetening and increases in aroma compounds so that ripening fruits are ready to eat or process. The associated physiological or biochemical changes are increased rate of respiration and ethylene production, loss of chlorophyll and continued expansion of cells and conversion of complex metabolites into simple molecules.

When fruits on a tree grow to its intended size and shape – *maturity* then, within a week or so, it *ripens* with visible changes in the following qualities:

1. *Aroma*. Bitter and astringent phenols fade away (their job was to discourage animals before the seed was ready), and nice aromas are produced (to encourage animals). This normally only happens while fruit still attached to tree. Ethylene gas is the "ripening hormone" that coordinates the ripening.
2. *Sweetness*, in the form of sucrose or fructose. It can come as sweet sap while attached to the tree or in some fruits by converting the fruit's stores of starch/glucose/acid.
3. *Juiciness* and *softness*: An enzyme polygalacturonase attacks pectin in the cell walls making cells slide around (softness) and spill their contents (juiciness). Acids are used up in this, making the fruit less sour.
4. *Colour* changes, brightens, forms a waxy sheen to slow loss of water. Look at the background colour, not the red blush which growers have bred so it appears even before ripeness.

Maturity and ripeness

There is a distinction between maturity and ripeness of a fruit. Maturity is the condition when the fruit is ready for eating or if picked will be ready for eating after further ripening. Ripeness is that optimum condition when colour, flavour and texture have developed to their peak.

Some fruits are picked when matured but not yet ripe. Since many fruits ripen off the tree, unless they were to be processed quickly, some will become overripe before they could be utilized if picked at peak ripeness.

Senescence

Senescence can be defined as the final phase in the ontogeny of the plant organ during which a series of essentially irreversible events occur which ultimately leads to cellular breakdown and death.

Death

Death is an irrecoverable state or occurrence of fruit in which both physiological and bio-chemical activities of cells that make up a fruit has ceased to function. At this stage the fruit is said to have deteriorated.

2.4 Fruits and vegetable harvest technologies

Fruits are easily damaged at harvest time, so care is required. The selection of a harvesting procedure will depend on the characteristics of the product. For commodities destined for the fresh market, integrity and appearance are important. This is especially true for commodities such as lettuce, berries, grapes, peppers, apples etc. that can be damaged easily. With manual harvesting, personal hygiene is particularly important since there is a great deal of handling that could lead to product contamination.

2.5 Time of harvest factors

Time to harvest horticultural products is of essence in determining the quality of such products. Proper time to harvest depends upon several factor which include variety, location, weather, ease of removal from the tree and purpose to which the fruit will be put. For example, Orange changes with respect to both sugar and acid content as then ripen. Sugar level increases while acidic content decreases. The ratio of sugar to acid determines the taste and acceptability of the fruit and the juice. Fruits to be canned are picked before it is fully ripe for eating since canning will further soften the fruit.

Measuring internal quality of fruits prior to harvest is vital to ensure that fruit reaches the market with good eating qualities acceptable to most consumers. Three internal maturity parameters must be measured to determine the best time to harvest fruit. These are the juice content, total soluble solids (percentage sugar or °Brix) and citric acid content. When these parameters have been measured, then the sugar-acid ratio is then calculated.

Collecting fruit samples

When collecting fruit for testing, it is important that the sample represents what you will be harvesting, that is, the same size/colour. If you are strip picking, remember to sample all sides of the tree as well as fruit from inside the canopy.

Figure 2-7: Weighing sample fruits

It is important to remember variations in your farm, such as topography and soil type, when collecting a sample, as they may have an effect on ripening. Collect a minimum of 10 fruit per variety to be tested. The more fruit tested the more accurate the test is.

- Weigh sample and record results; tare or zero the scales when using a tray.
- Weigh your 10 fruit and record the combined weight in grams.
- Weigh the empty 1 or 2-litre jug and record the weight in grams.
- Juice all 10 fruit using the juicer. Apply even force and try to remove all the juice.
- Strain the juice into the jug. Weigh the juice and record the weight in grams, then subtract the weight of the jug.
- Calculate percentage juice content by dividing the juice weight by the total fruit weight. Multiply this by 100 to get the percentage.

Procedures for measuring these parameters are described below.

Step 1: Calculating juice content - percentage juice

Juice content is an important parameter for the measurement of internal quality of fruits. Under- or over-ripe fruit tends to be less juicy, which directly affects eating quality.

If the total weight of fruit samples collected is 600 grams, and the total juice weight extracted from the samples is 300 grams, the percentage juice in the fruit is calculated thus:

$$Percentage\ juice = \frac{Juice\ weight}{Fruit\ weight}\ x\ 100 \ldots\ldots\ldots\ldots\ldots\ldots\ldots 2.2$$

For the example above,

$$Percentage\ juice\ =\ 300 \div 600\ x\ 100\ =\ 50\%$$

Remember to subtract the weight of the jug from the weight of the juice and jug. Alternatively place the empty jug on the scales and zero the scales prior to adding the juice.

Step 2: Calculating total soluble solids - percentage sugar or degrees Brix

Sugar levels are a common measurement in a wide range of crops. As the fruit grows and flesh is formed, it deposits nutrients as starch which later transforms to sugars as the fruit ripens. The percentage sugar, measured in degrees Brix (°Brix) using a refractometer, indicates the sweetness of the fruit by measuring the soluble solids in the juice.

To use the refractometers, follow these procedures:

- Direct the refractometer to a good light source.
- Ensure the refractometer prism surface (glass) is clean and dry.
- Place a small amount of the fresh juice (a couple of drops is sufficient) onto the prism.

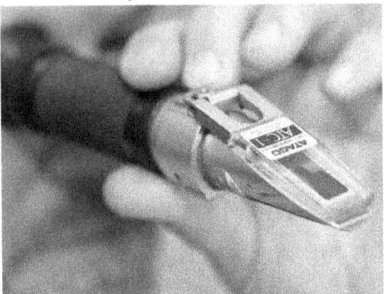

Figure 2-8: Reading from refractometer

- Look through the eyepiece (Figure 2-8) while pointing the prism in the direction of good light (caution: not directly at the sun).
- Focus the eyepiece and take the reading where the base of the blue colour sits on the scale — this reading is the sample's Brix.
- Clean the refractometer immediately with a damp tissue, and dry thoroughly.

Note: If you are using a digital refractometer place two drops of fresh juice on the prism (glass) and record the reading. Clean the prism immediately after use.

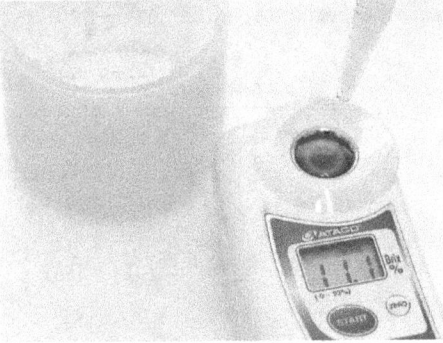

Figure 2-9: Digital refractometer

Step 3: Calculating citric acid content

Citric acid gives citrus its tartness and is in highest concentration early in the season, decreasing as the fruits mature. Citric acid content can be determined using two methods: the use of a chemical indicator called phenolphthalein and the use of a pH meter.

Both methods involve titration, which means adding a solution of known concentration to a solution of unknown concentration until a desired reaction is achieved.

Method 1: Using phenolphthalein

a. Draw 10 milliliters of juice into the pipette and transfer the juice to a clean conical flask. Clean the pipette immediately after transferring the juice.
b. Add two to three drops of phenolphthalein indicator to the conical flask and carefully swish mixture.
c. Squeeze the burette fill bottle, containing (0.1) normal sodium hydroxide (NaOH).
d. Open the burette tap and allow a trickle of NaOH to run into an empty beaker. This is to ensure no air is trapped in the burette prior to use.
e. Squeeze the burette fill bottle again until the NaOH level in the burette reads zero at the top of the scale.
f. Hold the conical flask containing the juice mix under the burette and while swirling the flask slowly add the NaOH to the juice by opening the burette tap.
g. Keep swirling the flask while adding NaOH, until the solution just starts to turn pink. As soon as the solution turns pink close the burette tap. Note the swirl of pink in the sample Look at the scale on the burette and record how much sodium hydroxide you have added to the flask. Multiply the figure you have recorded by 0.064 to give you the citric acid level of your sample.

The first few times you do this procedure, it may be difficult to see the colour-change point. If you look closely you will see the juice mix slowly lighten in colour, almost becoming clear and then change to a light green colour. This is the point just before the end of the test and a few extra drops of the sodium hydroxide will make the solution turn pink.

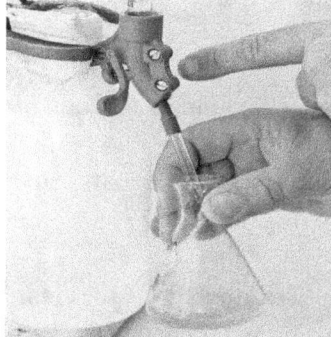

Figure 2-10: Adding the 1% Normal sodium hydroxide to the solution

If you go past this point and the solution changes from pink to a deep purple/orange, you have added too much sodium hydroxide and you will need to empty the conical flask and begin again.

Method 2: Use of a pH meter

a. Draw 10 milliliters of juice into the pipette and transfer to a clean beaker. Clean the pipette immediately after transferring the juice.

b. Carefully place pH meter probe into the solution and turn on (remember to calibrate the pH meter first if required).

c. Squeeze the burette fill bottle, containing 1% (0. 1) normal sodium hydroxide (NaOH), to fill the burette.

d. Open burette's tap and allow a trickle of NaOH to run into an empty beaker. This is to ensure no air is trapped in the burette prior to use.

e. Squeeze the burette fill bottle again until the NaOH level in the burette reads zero at the top of the scale.

f. Hold the beaker containing the juice mixture under the burette and while swirling the beaker, slowly add the NaOH to the juice.

Keep swirling the beaker whilst slowly adding NaOH to the solution until it reaches a pH of 8.2. Look at the scale on the burette and record how much sodium hydroxide you have added to the flask. Multiply the figure you have recorded by 0.064 — this is the citric acid level of your sample.

Example calculation

If the amount of NaOH added is 24.2 milliliters, the citric acid concentration (grams per 100 millilitres) is millilitres of NaOH x 0.064

$$Citric\ acid\ conc\ =\ 24.2\ x\ 0.064\ =\ 1.55\ grams\ per\ 100\ millilitres$$

Step 4: Calculating sugar-acid ratio

The sugar-acid ratio contributes to the unique flavour of citrus. At the beginning of the ripening process the sugar-acid ratio is low, because of low sugar content and high fruit acid content—this makes the fruit taste sour. During the ripening process the fruit acids are degraded, the sugar content increases and the sugar-acid ratio achieves a higher value.

Calculate the sugar-acid ratio by dividing the °Brix identified in Step 2 by the citric acid concentration identified in Step 3. This will give you the sugar-acid ratio of your sample.

Example calculation (sugar-acid ratio)

If the sugar concentration equals 15.2°Brix and citric acid concentration is 1.55g per 100mL,

$$Sugar - acid\ ratio\ =\ \frac{°Brix}{Citric\ acid\ concentration} \dots\dots\dots\dots\dots\dots\dots. 2.3$$

$$Sugar - acid\ ratio\ = \frac{15.2}{1.55} = 9.8$$

2.6   Harvest time (picking time) and harvest handling

The best time to harvest products is during the coolest part of day, which is usually in the early morning and late evenings, to maintain low product respiration. Produce should be kept shaded in the field. Immediately after harvest, products should be placed in a well-ventilated shady environment to prevent high temperature increases that occur if products are exposed to direct sunlight. For some commodities, such as berries, tender greens, and leafy herbs, one hour in the sun is too long. Shade the harvested product in the field by covering harvest bins with reflective pad to reduce heat gain from the sun, water loss, and premature senescence.

Do not compromise high quality product by mingling it with damaged, decayed, or decay-prone product in a bulk or packed unit. Avoid unnecessary mechanical damage; bruising, crushing, or damage from humans, equipment, or harvest containers. If possible, move the harvested product into a cold storage facility or postharvest cooling treatment as soon as possible. Rapid cooling to reduce respiration rate is used commonly in the horticultural industry, particularly for highly perishable products such as strawberries and apples. Hydro cooling, vacuum cooling, or forced air-cooling is used widely to remove field heat rapidly. Use clean and sanitized packing or transport containers.

2.7   Fruit harvest technologies

Two technologies are popularly employed in fruits and vegetables harvest; Manual harvest technology and mechanical harvest technologies.

2.7.1 Hand picking technology

In some countries fruit is harvested by hand, placed onto straw on the ground under the trees, hand-sorted, and packed into containers before leaving the orchard. Under these conditions consistent quality control is difficult to achieve among orchards, but fruit may sustain less handling damage. Fruit are easily damaged at harvest time, so care is required.

Figure 2-11: Picking mango fruit

Manual harvest is usually accomplished with the use of *hand* or *mechanical* aids. Hand harvesting is widely used for products or commodities destined for the fresh market.

A correct hand harvesting includes some simple but essential rules:

1. Place picked fruits in basket carefully to avoid mechanical damage
2. Clean the harvesting basket and your hand thorough.
3. Pick fruit (always) when it is ready for processing.

Note that proximity of the processing center to the source of supply for fresh raw materials present major advantages such as:

1. Possibility of picking the best suitable fruit
2. Reduction losses by handing and transportation
3. Ministries transport cost
4. Possibility of using simpler receptacles for raw material transport.

Mechanical aids are available for harvesting in the form of gantries, picking ladders, and in some cases mobile conveyor systems, but more commonly the picker collects the fruit into a small holder such as a picking apron or bucket.

Mangoes, papaya, apples, and fruit grown in tall trees can be harvested using picking poles: Fruit is separated from the tree by a sharp cutting edge on the end of the pole and falls into a net just under the cutter. In some countries fruit is harvested by hand, placed onto straw on the ground under the trees, hand-sorted, and packed into containers before leaving the orchard.

Fruit picking tools

Some fruits need to be clipped or cut from the parent plant. Clippers or knives should be kept well sharpened. Peduncles, woody stems or spurs should be trimmed as close as possible to prevent fruit from damaging neighbouring fruits during transport. Pruning shears are often used for harvesting fruits, some vegetables, and cut flowers. Varieties of styles are available as hand held or pole models, including shears that cut and hold onto the stem of the cut product. This feature allows the picker to harvest without a catching bag and without dropping fruits.

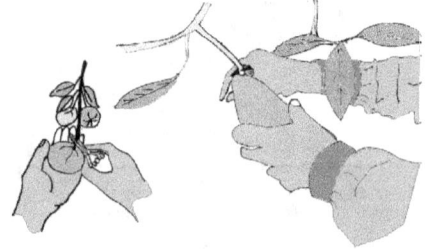

Figure 2-12: Harvesting fruits with tools

Fruit trees are sometimes quite tall and letting fruit fall to the ground when it is cut from the tree which could cause severe bruising. If two pickers work together, one can clip or cut the fruit from the tree, and the other can use a sack or canvass to break its fall. The catcher supports the bag or canvass with his hands and other end tied to the tree, catches the falling fruit, then lowers the far end of the bag or canvass to allow the fruit to roll safely to the ground.

Figure 2-13: Plucking fruits on canvas (catching surface) from tall tree

Harvesting containers

Picking baskets, bags and buckets come in many sizes and shapes. Buckets are better at baskets in protecting produce, since they do not collapse and squeeze produce. These harvesting containers can be made by sewing bags with openings on both ends, fitting fabric over the open bottom of ready-made baskets, fitting bags with adjustable harnesses, or by simply adding some carrying straps to a small basket.

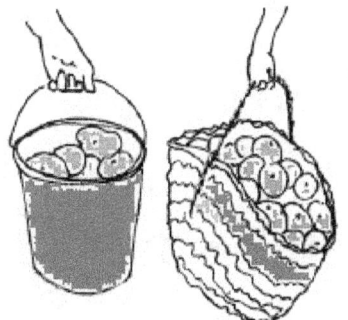

Figure 2-14: Harvest containers

Advantages and disadvantages of manual harvesting

The primary advantages manual harvesting includes:

a. Harvesting of fruit or vegetable can be done at appropriate maturity.
b. The produce will suffer minimum damage.

Disadvantages includes

a. It is a time consuming process.
b. More labour is required during harvesting season.

2.7.2 Mechanical method of fruit and vegetable harvest

Mechanical harvesting is recommended for produce that can readily withstand physical handling e.g., carrots, potatoes and radishes etc). It is generally used to harvest produce destined for the processing industry.

Mechanical harvesters have been developed for many crops including apples, strawberries, blackcurrants, blueberries, cherries, and raspberries. Harvesting involves shaking the tree or using long stick to cause mechanical vibration and catching the detached fruit underneath in a large blanket or net or by the use of mechanical aids such as harvesting tools and mechanical harvesters.

Tree shaking: Harvesting involves shaking the tree or cane by mechanical vibration and catching the detached fruit underneath in a large blanket or net. However, these systems can cause significant damage to the crop and are generally only suitable for fruit to be used for processing. There are difficulties if the fruit on the tree do not all ripen at the same time.

Mechanical aids: Mechanical aids are improvements on the tree shaking and hand picking methods of harvesting fruits. Mechanical aids are available for harvesting in the form of gantries, picking ladders, and in some cases mobile conveyor systems, but more commonly the picker collects the fruit into a small holder such as a picking apron or bucket, which holds not more than 15 kg of fruit. Mangoes, papaya, apples, and fruit grown in tall trees can be harvested using picking poles. The fruit is separated from the tree by a sharp cutting edge on the end of the pole and falls into a net just under the cutter.

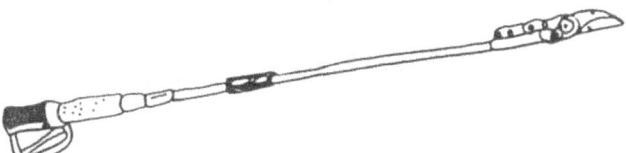

Figure 2-15: Picking pole

Using a cutting tool attached to a long pole can aid picking of crops such as mangoes and avocados when the fruit is difficult to reach. Cutting edges should be kept sharpened and the catching bag should be relatively small The angle of the cutting edge and the shape of the catching bag can affect the quality of the fruit harvested, so it is important to check performance carefully before using any new tools.

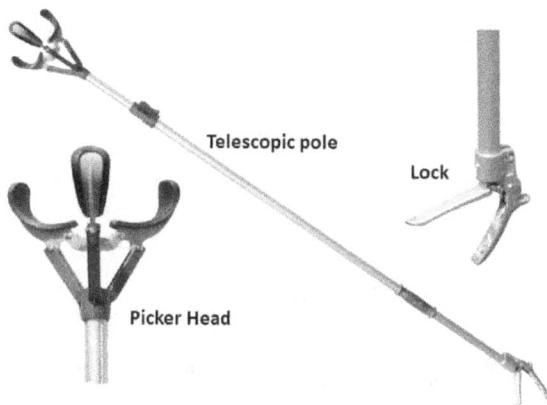

Figure 2-16: Tele poles for fruit picking

Mechanical harvesters: Mechanical harvesters have been developed for many crops including apples, strawberries, blackcurrants, blueberries, cherries, and raspberries. Some fresh vegetables are harvested by mechanical means. These include peas, beans, tomatoes, sprouts, and root crops, particularly if the product is for processing.

Figure 2-17: Trunk shaker/deflector fruit harvester

Advantages and disadvantages of mechanical harvesting

The primary advantages manual harvesting includes:

a. The produce can be harvested at a faster rate.
b. Less manpower is required as compared to hand harvesting.

Disadvantages

a. Damage can occur to crops.
b. Not suitable for marketing of fresh commodities

Effects of mechanical harvesting methods

Damage during mechanical harvesting can lead to a number of undesirable changes in produce. Physical damage caused by mechanical harvesting methods may lead to:

1. Water loss
2. Increased respiration rate
3. Initiation of ethylene synthesis
4. Production of undesirable colours (browning)
5. Penetration of plant tissues by micro-organisms (both plant and food borne human pathogens)

2.8 Harvest effect on postharvest quality

Great care must be taken during harvesting of perishable fruits and vegetables to avoid physical damage. Any mechanical damage that occurs at harvest, during movement of product to the pack house, or through grading and packing lines will result in enhanced respiration, increased ethylene production, water loss and increased susceptibility to infection by postharvest pathogens, all of which can induce rapid deterioration and loss of quality.

A number of simple but effective steps can be taken to reduce physical damage from occurring during this phase of the harvesting and handling system. These include careful handling of the product at all stages of the operation, good sanitation and hygiene with all equipment, maintenance of packing equipment to prevent excessive drops onto hard surfaces, and padding of all machinery surfaces on which products may impact.

2.9 Mechanical damage

The mechanization of various harvesting and subsequent manipulation operations has an unfavourable consequence on harvested fruits in that it leads to an increase in damage to the material processed. The quality of the product will be lowered and in many cases followed by rapid spoiling and complete deterioration.

Causes of mechanical damage

Damage to agricultural products may appear in different forms. The form of damage depends on the physical and biological construction of the product and on the type of load. Mechanical picking of fruit implies significant mechanical damage. Two major actions that cause damage in fruits include:

1. *Impact action*: When shacking a tree the fruit impacts against the tree branches, against other fruits and finally against the catching surface. The tissue beneath the skin is deformed by impact. If the deformation passes the biological yield point, tissue will turn brown within a short time and be spoilt. In certain cases, browning under the skin is not visible from outside (e.g. in pears). Certain fruits e.g. cherries, and sour cherries fall without their stems on shacking the tree. Where the stem is torn out of the fruit juice

appears, representing a loss on the one hand and promoting deterioration on the other. Fruit collected this way may be suitable for fast processing.

2. *Loading and unloading action*: Significant damage must be counted upon during loading and unloading from the means of transport. The main damage resulted from compression forces, dynamic forces and internal forces. The extent of damage also depends on the species, on the stage of ripening and on temperature.

It is therefore obvious that damage is generally caused by static and dynamic external forces. Mechanical damage resulting from internal forces is caused by physical variations taking place inside a product, for example variations of the temperature and moisture content, or chemical and biological variations. Cherries and tomatoes have skin cracking due to an increase in internal pressure.

Establishing and measuring damage

The methods of assessment established in practice are described below, from which the most appropriate to given circumstance must be selected. The main forms of appearance of mechanical damage are as follows:

Abrasion: Superficial damage produced by any type of friction (other fruits, packaging materials, packing belts, etc.) against thin layer of the skin. The skin is damaged or partly separated from the tissues beneath. Abrasion is sometimes hardly visible after harvesting but will become apparent after few days of storage.

Figure 2-18: Pear fruit abrasion

Bruise: Damage to plant tissue occurs as a result of external forces causing physical changes. Fruits and vegetables skin acts differently depending upon how fast you apply loads to them because of its visco-elastic property. Under slow loading, they act more rubbery and tough; under fast loading, they are more brittle. Dropping potato tubers, for example, from a low bruising height may cause one type of bruising (black spot discoloration with no cracking), while dropping the same tubers from a much higher height causes a very different type of bruising (cracks and shattering with no discoloration

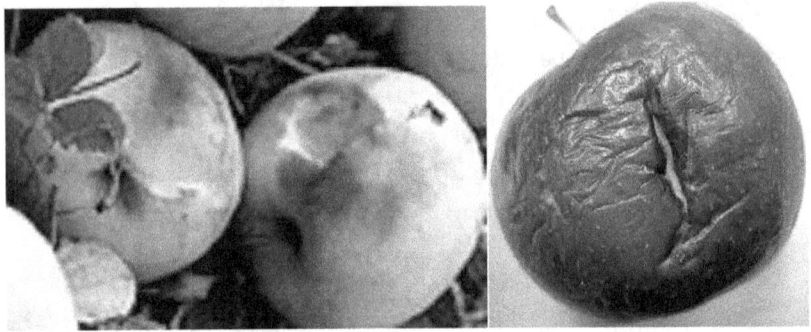

Figure 2-19: Pear fruit bruising

Crack: This is limited to cracking of skin or tissue due to impact or increase in internal osmotic pressure, without causing the product to fall apart into several pieces. Cracks appear weathered and corky if they occur in green fruit. Fruit may develop any of a number of fungal fruit rots at the point of cracking. Cracking could be *shatter cracking: A* multiple cracks starting radially from the point of impact or out from the stem scar; *Skin cracking:* Cracks restricted to the outer skin alone (Figure 2-20).

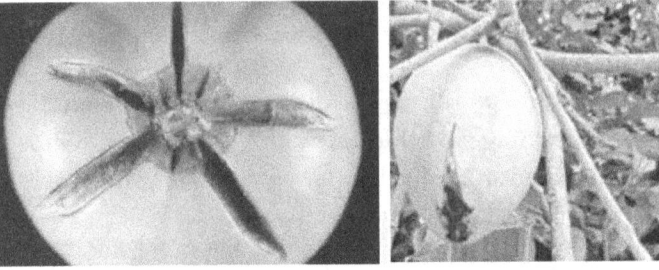

Figure 2-20: Forms of fruit crack

Cut: This is the penetration of a sharp tool into the product without any significant crushing effects.

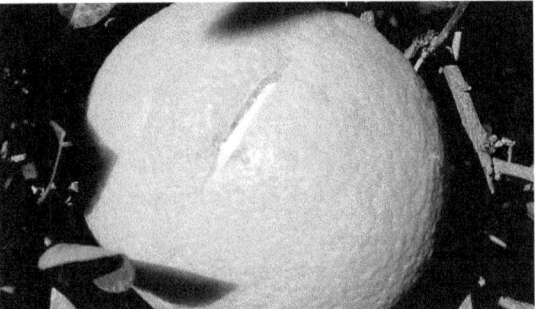

Figure 2-21: Orange fruit cut

Puncture: Caused by pointed needle-type tools, plants stems, or thorns penetrating the surface of a product and the tissue beneath.

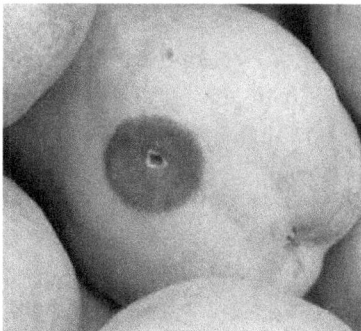

Figure 2-22: Fruit puncture

Split: A product divide into several parts Splitting is caused by an erratic watering pattern, particularly the excessive accumulation of water. The inside of the fruit is under great pressure and the fruit splits open when lightly tapped. The condition is weather related and there probably is a cultivar interaction as well. In Figure 2-23, notice that you can see entirely through this donut peach with a split pit. This trait is undesirable and would make the fruit unmarketable.

Figure 2-23: Fruit picking heads

Tear: fruit tear is usually caused by stem ends i.e. when the skin of a fruit is thorn on removing the stem. Donut peaches (Figure 2-24) often have a very short peduncle (stem) and if you are not careful in picking them, the skin may tear at the point where the fruit attaches to the stem.

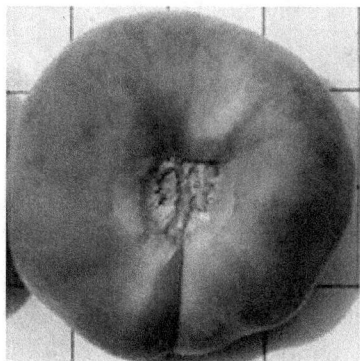

Figure 2-24: Fruit tearing at stem end while picking

Distortion: Distortion is simply deformation or change of form of products under pressure caused by loads acting on it. This often occurs during storage and bulk transportation and is caused by the weight of the mass of fruits on bottom layers. It also happens when the packed mass exceeds the volume of the container (Figure 2-25) or by the collapse of weak boxes or packages unable to withstand the weight of those piled up high.

Figure 2-25: Compression injury in tomato

Factors affecting sensitivity to damage

Sensitivity to damage is influenced by numerous factors. One groups of parameters concern the physical and biological state of the material e.g. temperature, moisture content, stage of growth, and ripeness, while others are related to the load characteristics e.g. static, dynamic, oscillating, loading etc.

In most cases, temperature greatly effects the mechanical properties of agricultural products and thereby their sensitivity to damage. With variation of temperature, the tugor pressure of cellular material and together with it the elasticity both vary.

Good produce quality is attributable to appropriate production practices, careful harvesting, and proper packaging, storage, and transport. Handling is the final stage in the production process of high quality fresh product. Being able to maintain a level of freshness from the field to the dinner table presents many challenges.

Bruises and other mechanical damage do not only affect appearance, but provide entrance to decay organisms as well. Handle produce gently, and crops destined for storage should be free as much as possible from skin break, bruises, spots, rots, decay, and other deterioration.

2.10 Chilling injury

Chilling injury occurs when sensitive crops are exposed to low temperatures that are above freezing point. These chilling temperatures cause break down of cellular membranes, resulting in loss of compartmentalization within the cells of the tissue, increased leakiness, water soaking of tissue, and eventually pitting or browning. Different parts of some vegetables have distinct sensitivities. In eggplant, the cap or calyx is more sensitive and turns black before the fruit itself is affected. Some chilled fruit fail to ripen normally, while in others there is an accelerated rate of senescence and a shortened shelf life.

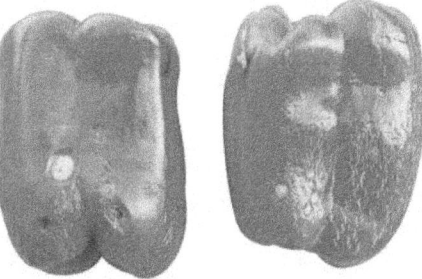

Figure 2-26: Symptoms of chilling injury in bell pepper

Most tropical and subtropical products (like tomato, eggplant, green beans, okra, and banana) are susceptible to chilling injury when exposed to temperatures above freezing but below a critical threshold temperature for each particular product, while others are injured by low temperatures and will store best at 45 to 55 degrees F. Damage often is induced by a very brief exposure, but may not become apparent for several days or until transfer to warmer conditions.

Figure 2-27 shows a Cavendish Williams bananas harvested at the hard green stage from the same banana branch stored at 22°C for 11 d (non-chilled) and placed at 4°C for 7 d (chilled) before transfer to 22°C for 4 d. Comparing the bananas they ripened, the chilled bananas failed to yellow and instead developed extensive peel blackening due to cell death. The non-chilled bananas gradually turned from green to yellow.

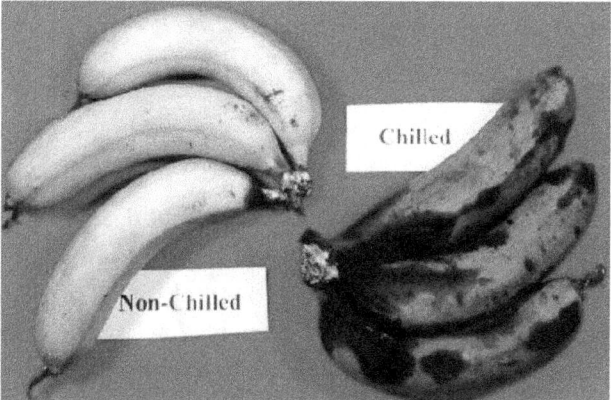

Figure 2-27: Chilling injury in bananas

Crops' sensitivity to chilling injury

Crops such as cucumbers, eggplants, pumpkins, banana, okra, and sweet potatoes are highly sensitive to chilling injury. Moderately sensitive crops are snap beans, muskmelons, peppers, winter squash, tomatoes, and watermelons. These crops may look sound when removed from low temperature storage, but after a few days of warmer temperatures, chilling symptoms become evident.

Chilling injury factors

Both time and temperature are involved in chilling injury. Damage may occur in a short time temperatures are considerably below the danger threshold, but some crops can withstand temperatures a few degrees into the danger zone for a longer time. The effects of chilling injury are cumulative in some crops. Low temperatures in transit, or even in the field shortly before harvest, add to the total effects of chilling that might occur in storage.

Effects/ symptoms of chilling injury

The effects of chilling injury are cumulative in some crops. Chilling injury may not be apparent until produce is removed from low-temperature storage. Depending on the duration and severity of chilling, chilling symptoms become evident in the following ways several hours or a few days after the produce is returned to warmer temperatures:

⇨ Surface pitting and localized water loss
⇨ Surface browning or other skin blemishes
⇨ Internal discolouration to vascular tissue or in parenchyma cells
⇨ Increased susceptibility to decay, water soaking
⇨ Failure to ripen or uneven colour development
⇨ Loss of flavor, especially characteristic volatiles
⇨ Development of off-flavours
⇨ Meatiness or wooliness of texture and
⇨ failure to ripen among several others

Methods of reducing chilling injury

Increasing interest is being shown in the potential of short-term pre storage exposure to relatively high temperatures (35–55 °C) as a means of reducing chilling injury, as a quarantine treatment for disinfesting products of pests and for reducing the rate of some senescence processes. Product-specific high-temperature treatments have been shown to reduce ethylene production, prevent yellowing and retard softening in a number of products, all of which can extend storage life.

CHAPTER 3

FRUITS & VEGETABLE QUALITY MEASUREMENT

Content: Introduction to quality and quality measurement, quality change prediction, determination of quality, quality and postharvest changes quality assurance, and product contamination

3.   Introduction

Quality is an important factor in the production and marketing of biological and agricultural products. Quality plays a prominent role in ensuring the competitiveness and sustainability of fruit and vegetable products in the open market.

Vegetable quality is of paramount importance in order to sustain consumer demand and to be competitive in the global market. There is a need to introduce the concept of a quality assurance system in the handling system in order to deliver high quality vegetables for the local and export markets.

Quality assessment is very subjective and depends on a person's viewpoint and preferences. There are three key areas to be considered in quality assessment: meaning of quality, quality measurement, and maintenance of quality. Let's look at each of these items as discussed below:

3.1   Meaning of quality

In the ISO 9000 standard (developed by the International Standards Organization), quality is defined as "the totality of features and characteristics of a product or service that bear on its ability to satisfy stated or implied needs". For fruit and vegetables it is "the combination of attributes or properties that give them value in terms of human food." Quality is perceived differently depending on the needs of the particular customer. The customer purchasing the fruit for consumption usually judges the product on the basis of its appearance, shape, firmness, and colour, as well as freedom from defects such as spots, marks, or rots.

The consumer will judge the fruit on its eating quality (i.e., taste or texture) as well as its keeping qualities in the home. If something is not a quality product, this implies that the product does not meet a certain standard that has been adopted by the consumer. In this case, the market price is adversely affected.

3.2   Quality needs of fruit and vegetable

It is becoming increasingly important to present top-quality products to consumers. Most major supermarket chains buy horticultural products to specifications (size, weight, colour, and freedom from defects); buyers are rejecting products that deviate from pre-contracted specification ranges.

It is therefore imperative that growers, exporters, shippers, and scientists work together to ensure that customer quality standards are being met. An important first step in this process is to ensure that all concerned individual understand the nature of product perishability and appreciate the important biological factors that influence deterioration.

3.3   Quality change prediction and maintenance in fruit and vegetable

Quality change prediction

The criterion by which the acceptability of a food is judged is that of quality. Quality is a complex function of many attributes of a food which include flavour, texture firmness, size shape and appearance. Because of the difficulty in specifying quality, the concept of a quality indicator often becomes a tool used in quality change prediction.

Quality indicator is the application of one or more attribute of a product to indicate storage changes. The variability makes hypothesis based on it unpredictable. Because temperature is the major factor influencing quality, it is very desirable to be able to predict the quality change from a known time-temperature data as a function of storage time.

Quality maintenance in fruits and vegetable

All living things including fruits and vegetable respire to generate energy for continued metabolism. Respiration rate of horticultural products varies, but as a general rule perishability is a function of respiration rate; the greater the respiration rate the more perishable the product and the shorter the time it can be stored and still maintain acceptable quality. Actively growing products, such as asparagus, or tropical fruits, such as mango, have high rates of respiration and limited storage life.

Quality cannot be improved after harvest, but can only be maintained; therefore it is important to harvest fruits, vegetables, and flowers at the proper stages and sizes and at peak quality.

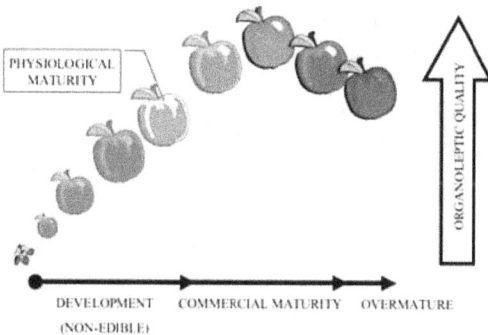

Figure 3-1: Organoleptic quality of a fruit in relationship to its ripening stage

3.4   Fruit quality indicators

Many quality measurements can be made before a fruit crop is picked in order to determine if proper maturity or degree of ripeness has developed. The following quality indicators help measures the quality of fruits before harvest.

1. *Colour* may be measured with instruments or by comparing the colour with standard pictures chart.
2. *Texture* may be measured by hand compression or by the single type of penetrometer plunger.
3. *Soluble solids content*: The concentration of soluble solids (sugar) in juice can be estimated with a refractometer or a hygrometer. The refractometer measures the ability of the solution to bend or refract a light beam which is proportional to the solution concentration.
4. *Juice density*: A hygrometer will float in the juice at a height relative to the juice density.
5. *Titratable acidity*: Acid concentration can be measured by a single chemical titration on the fruit juice.

3.5   Methods for determining quality of fresh commodities

The following methods are used in the determination of quality of fresh commodities: Visual examination, size and shape factor, colour code, gloss factor and extent of defect.

a. *Visual examination*: The visual appearance of fresh fruits and vegetables is one of the first quality determinants made by the buyer whether the wholesaler, retailer or consumer.

Often the appearance of the commodity is the most critical factor in the initial purchase (in addition to price) while subsequent purchases may be more related to texture and flavour.

b. *Shape and size factor:* Uniform and characteristic shapes are important quality characteristics. Misshapen products may be more susceptible to mechanical injury and are generally avoided by consumers. Another example where shape is important is for broccoli. For the fresh market, compact broccoli florets are desirable while for fresh-cut, space between the florets is important to allow for cutting without injury. Size of product can also be important depending on its intended use.

Consumers tend to associate large size with higher quality and view larger fruit as more mature. A subjective evaluation of size and shape can be conducted on incoming product once the desirable and undesirable characteristics are determined. Size and shape charts are available for various commodities and weight is a fairly accurate measure of product size. The percentage of product which does not meet the desired characteristics can be recorded.

c. *Extent of defects:* The product should be evaluated for the presence of defects. The level of tolerance for each type of defect such as cuts, bruises, disease, low-temperature injury, and physiological disorders should be determined. During quality evaluation, the percentage of fruit with each class of defect can be determined as a guide to overall product quality.

A scoring system such as indicated below can be used to describe the incidence and severity of defects.

Score	Severity of defects
1	None defect,
2	Slight defect,
3	Moderate defect,
4	Severe defect, and
5	Extreme defect

d. *Colour:* We perceive colour when light reflected off the fruit or vegetables' surface falls upon the eye's retina; there is no colour without light. Colour perception depends on the type and intensity of light, chemical and physical characteristics of the commodity, and the person's ability to characterize colour.

Evaluating colour can be subjective or objective:

Subjective evaluation

The human eye is used to evaluate colour. Determination of commodity colour can be accomplished subjectively through the use of comparative colour charts. Colour charts can be very effective and useful if the colours truly match the colour change in the commodity of interest.

Advantages:

a. Faster and easier than objective measures.
b. Requires no specialized equipment.
c. Colour charts or guides can be used as references for matching and describing colours as in bananas, nectarines and tomatoes.

Disadvantages:

a. Results can vary considerably due to human differences in colour perception and human error.
b. Available light quantity and quality can influence colour perception.

Subjective scoring of colour may be more practical and faster and values can be referenced to objective colour values and to pigment concentrations. For small leafy tissues, for example, samples representative of a 5 point colour scale are evaluated for objective colour values and chlorophyll and carotenoid concentrations. Routine evaluations are done by subjective scoring, but referencing to objective measurements adds valuable information to the scores.

Objective evaluation

Objectively, colour is determined with a Minolta colourimeter; an instrument is used to provide a specific colour value based on the amount of light reflected off the commodity surface or the light transmitted through the commodity. The Minolta colourimeter can detect small differences in colour and provides separate values for lightness to darkness, green to red and blue to yellow scales.

Advantages

a. Less variability in colour measurement.
b. Can measure small differences in colour accurately.
c. Can be automated on the packing line.
d. Portable hand-held units are available (Figure 3-2).

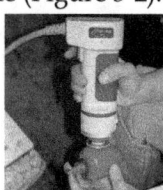

Figure 3-2: Minolta colourimeter

Disadvantages

a. Requires specialized equipment at a significant cost.
b. May be slower than subjective evaluation.

Colour code: Based on the UCLA center for human nutrition's colour code system, the table below showed types of fruits that exhibit the colour characteristics shown.

Table 3-1: Fruit colour code

Colour	Type of fruit/food
Red	Tomatoes, tomato products (pasta sauce, tomato soup, tomato-based juices, ketchup), pink grapefruit, watermelon
Red/Purple	Grapes, grape products (red wine, grape juice), prunes, cranberries, blueberries, blackberries, strawberries, red peppers, plums, cherries, eggplant, red beets, raisins, red apples, red pears
Orange/Yellow	Orange juice, oranges, tangerines, yellow grapefruit, lemon, lime, peaches, papaya, pineapple, nectarines.
Orange	Carrots, mangoes, apricots, cantaloupes, pumpkin, acorn squash, winter squash, sweet potatoes
Yellow/Green	Spinach, collard, mustard greens, turnip greens, yellow, corn, avocado, green peas, green beans, green peppers, yellow peppers, cucumber, kiwi, romaine lettuce, zucchini, honeydew melon, muskmelon
Green	Broccoli, brussels sprouts, cabbage, cauliflower, Chinese cabbage
White/Green	Garlic, onions, leeks, celery, asparagus, artichoke, endive, chives, mushrooms

Colour notation

The colour notations used in quality determination include:

a. *Hue:* Red, yellow, green, blue, purple or intermediate colours between adjacent pairs of these basic colours, e.g. RY, YG, GB, BP.

b. *Value of lightness:* The degree to which an object is judged to reflect more or less light than another object.

c. *Chroma or saturation:* The degree of departure from the gray of the same lightness.

d. *Gloss:* Gloss is a visual aspect of quality that depends on the ability of a surface to reflect light. Products that are freshly harvested often have a bright, glossy surface and this appearance factor can be greatly reduced with weight loss and other postharvest handling conditions. There are small portable instruments for gloss measurement such as refractometer.

Colour coded chart

Different fruits and vegetables get their colours from the various phytochemicals found in them, and those phytochemicals (chemical compounds that occur naturally in plants responsible for colour and other organoleptic properties, such as the deep purple of blueberries and the smell of garlic) offer different nutrients when eaten.

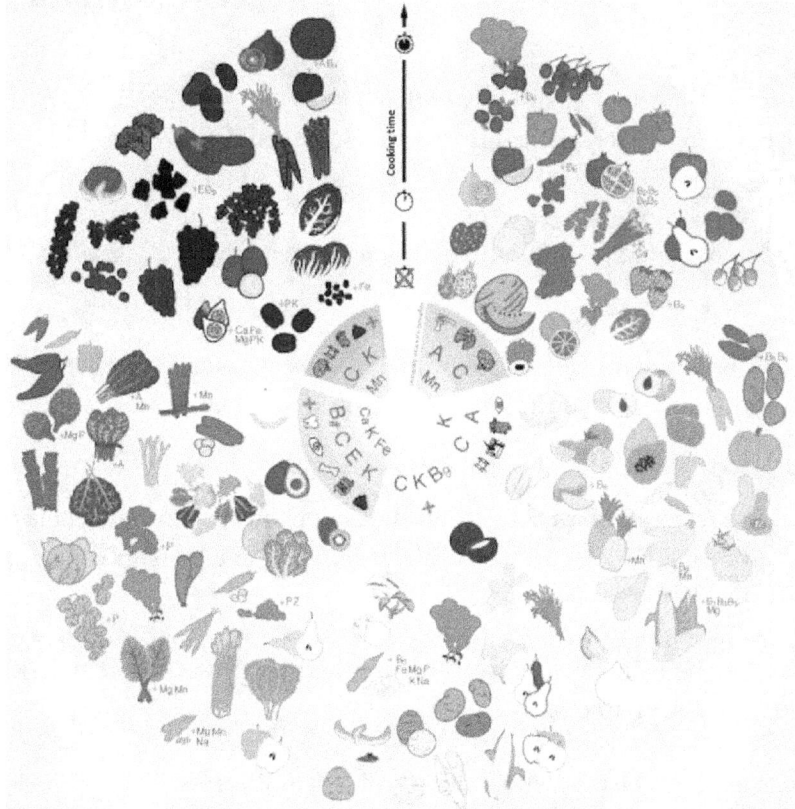

Figure 3-2: Colour coded chart

Colour-coded chart helps you pick the most nutritious fruits and vegetables for a diet. A colourful diet is a fast way to get a lot of vitamins and minerals without much effort. A quick glance at the chart shows you that to make a choice of foods that promote healthy bones and teeth for instance, leafy greens are a good option: lettuce, cabbage and brussel sprouts are all in the same category. They are also high in vitamins C and E, and a great source of iron.

If you love a good strawberry diet, you can find it in the chart in combination with other red fruits and vegetables, all of which promote good joint health and brain function, and are high in vitamins A, C and manganese. The chart is also divided up by required cooking time to get the nutrients involved, from low cooking time at the centre to longer times around the perimeter.

3.6 Factors affecting quality preservation of fresh commodities

The following factors play a major contributory factor in quality preservation of fresh commodities;

1. *Moisture content:* High moistures conditions lead to heat, insect and mould damage and are a major hazard to safe storage. Two moisture related factors that are important in drying and storage of agriculture product are equilibrium moisture content and safe storage moisture content.

2. *Temperature:* Extreme temperature variation in stored product could affect fruits and vegetables postharvest storage quality diversely

3. *Moisture migration during storage:* Properly harvested, dried and stored product will keep for a considerable length of time if conditions under which the grain is stored do not change.

4. *Mould growth:* Spoilage is likely to occur if no counter measures are taken. The lower limit for most mould growth is 40°f (4.4°c) and the optimum temperature for their growth 80-90°f (26-67-32°C).

5. *Respiration:* The respiratory rate of grain is dependent on the moisture content of the grain and it is questionable which part of the total respiration in a lot of grain is due to the fungi and which is due to the grain itself.

6. *Insect activities:* The activities of insects on postharvest preservation are such that active storage life of product reduces.

7. *Oxygen supply:* Much of the deterioration in high moisture stored products at ordinary ambient temperature is caused by respiration process which requires oxygen. Excluding oxygen from storage units causes aerobic respiration to cease.

3.7 Quality measurement of fresh commodities

The only true test of quality is the response of the consumer to acquisition of such product. Eating quality can be assessed most accurately by using taste panels. These consist of a selection of customers from the market who are trained to assess the quality attributes being examined. This is a very expensive and time-consuming task. The reliability and value of these methods depends on how well they correlate with the views of consumers, rather than the level of scientific objectivity of the test.

Measurements very often are given as indices, which can be used as guides to desirable quality attributes such as taste, texture, storage life, and maturity. In some cases visual observation is adequate; in others physical or chemical tests have been developed, often with limited scientific basis. These now serve as industry standards.

Measurement may be required at several different points in the postharvest chain. Particularly important points are when decisions must be made about the timing of operations. Therefore tests for harvest maturity are especially important. Many of these are crop-specific, based on particular physiological changes as the fruit or vegetable matures, and are no longer useful after harvest. Tests also are desirable at out-turn (unloading on arrival at the destination market), at retail, and in the home prior to consumption. In selecting a method the purpose for which the test is needed should be considered.

Measurement methods can be *destructive* or *nondestructive*. Destructive tests include tests that are themselves nondestructive but require samples to be cut from products. Tests can be further roughly grouped (according to the method used), into mechanical, visual (by human eye or by instrument), electrical, chemical, or biological. In some cases several methods may be combined to give an overall quality index.

3.7.1 Destructive methods

The following destructive methods of quality determination have been used in quality measurement in fruits.

Mechanical destructive indices

Firmness measurements in fruits and vegetables have been used for over 60 years as guides to the quality of the product. Firmness meters attempt to record a value that represents how easily the product can be deformed under a pressure applied to a limited area of its surface (or more simply what happens when you squeeze it). They range from laboratory systems to much cheaper handheld devices, which can be used in the field. Indenting heads include a penetrometer cylinder, a spherical indenter, and a flat plate to give compression tests between parallel platens. The following devices are widely used:

a. *Penetrometer:* The most common device used to assess firmness is the penetrometer. This has a cylindrical probe, the end of which is pushed into the object to be measured. The force required to give a predetermined penetration is recorded. In most fruit a section of the skin is removed first to expose the flesh.

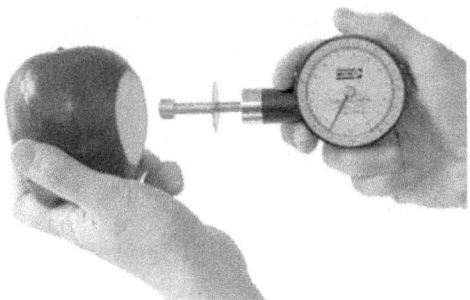

Figure 3-3: Effigi penetrometer

The penetrometer then gives an index of the firmness of the product tissue. Several versions have been developed. The Magness Taylor penetrometer is forced into the tissue, compressing a spring. When the probe has penetrated to a specified depth, the reading is taken from the spring compression. An alternative, more compact version is the Effigi penetrometer, which has a coiled spring so that the force can be read off a dial (Figure 3-4).

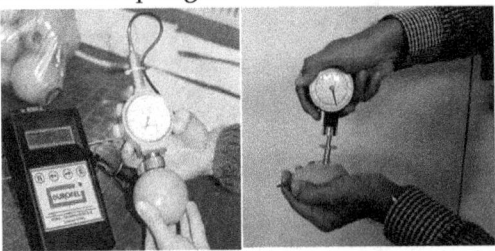

Figure 3-4: Using the penetrometer

The results from both these devices are quickly obtained, but values are very dependent upon the operator, to the extent that variations in readings of 200% have been recorded between different operators. The Effigi can be mounted easily into a drill press to reduce operator dependency.

The penetrometer still is used widely as an internationally recognized index for handling quality, storage life, and maturity in many fruit crops. The simplicity, speed of testing, and robustness of the equipment makes the penetrometer attractive to the fruit and vegetable industry. Moreover the test simulates the "feel" or firmness of the product. Cylinder heads are usually either 11 or 8 mm in diameter. However, it is not possible to interchange the heads, as the results from the two sizes are not consistent. In apples and pears the larger size is normally used, while the 8-mm probe is used on smaller fruit.

b. *Texture analyzers:* Research laboratories commonly use universal testing machines to determine physical properties of crops. Sensitive universal testing machines have been developed that are dedicated to the measurement of food texture. One system measures the bursting strength of the skin \ and the whole-fruit compression resistance in the same test. Measurements can be taken on whole fruits and vegetables or cut samples.

Figure 3-5: TAPlus high-performance food texture analyzer

Theoretical values can be obtained from stress–strain data, according to the loading geometry. Texture analyzers also can be used for nondestructive measurements.

c. *Twist tester:* An alternative destructive tester has been developed in which the fruit is pushed onto a blade mounted on a spindle, so that the blade enters the fruit at a predetermined depth under the skin. The fruit is rotated, so that the blade turns at a fixed depth (Figure 3-6).

A rising weight on the end of an arm is used to apply an increasing moment (torque) to the blade, to resist the rotation. Eventually failure of the tissue occurs, and the moment can be calculated from the angle of the arm. It has been used successfully to measure the texture of apples, mangoes, plums, and other fruits.

Figure 3-6: Twist tester

d. *Bruise and damage susceptibility testers:* Handling quality can be assessed using susceptibility testers to determine the susceptibility of fruit crop and vegetables to bruise.
e. *Abscission:* The point at which fruits and vegetables can be easily detached from their plants is essentially a nondestructive assessment of maturity, but because the measure cannot be undone, it is destructive. The test is simple, and in many crops, mature fruit are self-selecting by detaching them from the tree. However, in other crops abscission occurs too late, and their eating or storage quality is poor.

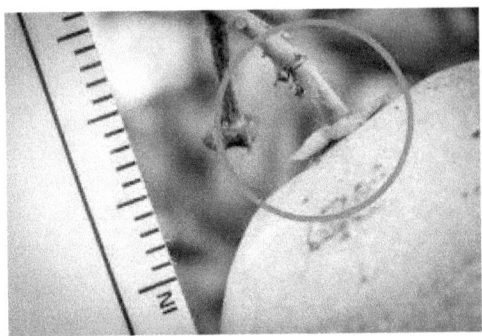

Figure 3-6: Point of abscission

Visual destructive tests

1. *Physiological changes:* Some indices require the visual assessment of cut sections of fruit and vegetables. Some fruits (e.g., bananas) change their cross-sectional shape noticeably as they approach maturity, and the cross-sectional area therefore can be used as an index [23]. Other fruits split open as they mature, to give a simple criteria for maturity. Pip, seed, or stone development inside the fruit also is often a good indicator of maturity.

2. *Colour changes:* Flesh colour is an important determinant of quality and maturity in many crops including melons, mangoes, and squash. Chemical analysis does not enable colour estimation: Vegetable and fruit yellowing is often a result of the disappearance of chlorophyll, which allows the yellow–orange xanthophylls and carotenes to become more visible. Blueberry colour is determined by anthocyanins, which are red when extracted.

   People can distinguish the level of lightness or colour intensity of an object, its hue (i.e., its colour name such as *red, blue,* or *green*), and its chroma (degree of colour purity, saturation, brightness, or greyness). Colour meters give an absolute determination of colour using a standard three-component specification, known as the Hunter Lab scale. In commercial practice colour charts normally are used. Carefully reproduced colour photographs of fruit in the desired state are common, although colour paint chips also can be used if the tissue is uniform in texture and colour.

3. *Level of soluble solid:* The levels of soluble solids in fruit and vegetable juices can be determined by measuring the refractive index of the juices. Laboratory and field devices require a small sample of juice placed on a glass cover. The refraction of the light produces an indication on a scale that gives a measure of the soluble solids directly. This is a useful indicator of maturity at harvest time. A refractometer is used to measure the soluble solid content of fruits.

Refractometer

Description: Refractometers have features such as automatic temperature compensation, thumb screw for easy calibration, Brix measurement range (e. g. measurements between 0-

32% ± 0.002 accuracy), and padded casing. Refractometers are low in cost but may require calibration. Measurements may be affected by temperature and delays in carrying out the test after exposing fresh juice.

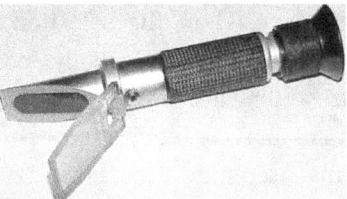

Figure 3-7: Refractometer

Uses: Handheld refractometer is used for measuring the amount of dissolved solids in fruits and vegetables. It is particularly useful for measuring sugar levels in grapes

Electrical destructive measurements

Electrical conductivity measurements have been used to determine moisture content of fruits. Electrical conductivity is a material's ability to conduct electric current. Some objects are much better conductors than others. Conductivity is measured by pressing two stainless steel or platinum probes into the product and applying a fixed or alternating voltage. A range of moisture meters is used for grain-moisture assessment.

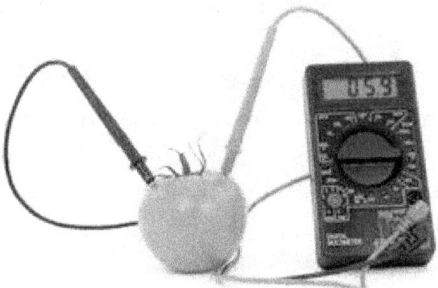

Figure 3-8: Electrical conductivity meter

3.7.2 Nondestructive methods

The following tests are in principle nondestructive, although in some cases a test may result in damage. For example, mechanical test pressures may exceed tissue failure strength in some samples, especially as the fruit softens. Care should be taken to ensure that the test is not producing any effect that would be harmful in the long term.

Visual nondestructive measurements

1. *Size and shape:* As fruits grow their shapes and sizes change. Assessment of size is one of the simplest methods for assessing maturity prior to and at harvest for many vegetables.

Mango maturity can be assessed in some cultivars by examining the position and angle the shoulder makes with the stalk and its point of attachment to the fruit

2. *Colour:* Surface colour is used widely for maturity and quality assessment and is probably the most common characteristic used in the selection and harvesting of many fruit. Generally assessments are still performed manually, particularly in the field; although inside packing sheds colour sorters are now available commercially for many crops. Although manual measurement may be subject to operator fatigue, human error, and variability, automatic systems are still often considerably inferior, because humans provide a more subjective assessment than machines.

   Colour assessment often can be complex: For example, in apples that develop a partial red colour, harvest maturity is based on the background colour, which remains in the green-to-yellow range of colours, even though the red colour appears to dominate. Human graders have been shown to fail to grade fruit correctly due to the influence of the blush colour. Machine vision systems are likely to continue to advance rapidly in speed and accuracy and should become increasingly reliable for discrimination of fruit maturity.

3. *Near-infrared reflectance and other electromagnetic radiation:* Near-infrared reflectance has been used to detect bruising in apples. Gamma-and x-rays also have been used to detect internal disorders in fruit and vegetables. Radiographic systems have detected water core in apples, hollow heart in potatoes, split pits in peaches, and the maturity of lettuce heads at harvest.

4. *Delayed light emission and transmittance:* Some fruit will re-emit radiation for a short time after exposure to a bright light. The amount of delayed radiation is a measure of the chlorophyll present, and this is inversely dependent on the maturity.

5. *Other visual methods of assessment:* A wide number of visual indicators can be used to determine maturity and quality. The state of foliage is particularly useful for many vegetable crops (e.g., optimum storage of potatoes may be obtained if they are dug up after the leaves and stems have died down)

Nondestructive chemical testing

Chemical methods of nondestructive testing include:

1. *Respiration rate:* Some fruits and vegetables change their rate of respiration as they mature and ripen. These changes can be used to identify when fruit and vegetables are near the end of their storage life and should be sold.

2. *Head gases and aroma:* Ripeness of fruit can be indicated by changes in the production of ethylene, or other volatiles. Although these changes may be too small for humans to detect, fruit flies and other insects may be attracted by volatiles.

3. *Biological:* Harvest dates can be estimated from the elapsed time after flowering for some fruits and vegetables. In practice it is difficult to be precise. Degree-day calculations

(cumulative product of time and temperature) are a guide to maturity for some greenhouse crops.

4. *Biosensors:* Biosensors that mimic taste and smell offer useful potential for quality assessment. These include ion-selective glass electrodes. These have enzyme or antibody film coverings that affect the number of hydrogen ions sensed. Metal-oxide gas detectors can react to specific gases. These detectors are still in the experimental stage.

Advantages of nondestructive testing methods

Nondestructive methods offer the following significant advantages over destructive methods.

1. There is a saving in the number of fruits or vegetables wasted,
2. The same fruit can be retested several times throughout its lifetime; this improves predictions of storage life.
3. Samples taken from packed fruit do not need to be replaced- a major advantage for quality-control inspection procedures
4. There is no mess or problem of disposal of sampled fruit—they can be repacked or returned to the packing line.
5. On-line assessment of every fruit is a possibility.
6. The number of samples required can be reduced because the same fruit can be used again and again without interrupting its normal life cycle,.
7. Tests that could be conducted in situ in the orchard
8. It would be possible to test fruit during growth, without removal from the tree, and to evaluate the relationship between orchard management and postharvest properties in a rigorous fashion.

Giving a reduction in variability as a result of random sampling of fruit in growth and storage trials (each fruit becomes its own control). This means that test procedures should become more reliable and rigorous, because measurements can be correlated better with fruit performance by tracking the development and storage behaviour of individual fruit.

3.8   Fruit firmness and its measurement

Firmness, or the degree of softness or crispness, is often measured using objective instruments. Subjective measure of firmness with the fingers can be useful for quick measures of gross differences in firmness, particularly of soft products.

Sample selection for firmness

The following steps are taken when selecting samples for firmness test:

1. Select a random sample of product from several representative boxes including at least 15 to 25 fruits or vegetables or 3% of the sample.
2. Select product with a uniform size to avoid variation in firmness due to size (large fruit are usually softer than smaller fruit).
3. Make sure all fruit tested are comparable in temperature since warm fruit are usually softer than cold fruit.

Proper use of firmness testers

4   Make two puncture tests per fruit (except very small fruit), once on each opposite cheek, midway between the stem and blossom end on sun and shade sides; avoid sunburned areas.
5   Remove a disc (larger than the tip to be used) of the skin with a vegetable peeler or sharp knife.
6   Use an appropriate tip (plunger), see Table 1, for each commodity.
7   All determinations for a given lot should be made by one person to minimize variability.
8   Hold the fruit against a stationary, hard surface and force the tip into the fruit at a uniform speed (it takes 2 seconds).
9   Depth of penetration should be consistently to the scribed line on the tip.
10 Record reading to the nearest 0.5 lb-force or 0.25 kg-force.

The table 3-2 below shows the recommended tip sizes for firmness test in some common fruits:

Table 3-2: Recommended tip sizes for firmness measurements.

Tip size	Commodities
11 mm (7/16inch)	Apple
8mm (5/16 inch)	Apricot, avocado, mango, nectarine, papaya, peach
3mm (1/8 inch)	Cherry, grape, strawberry
1.5mm (1/16 inch)	Olive

Proper units for firmness testers

It is inappropriate to use the term "pressure" in association with firmness measurements using the devices described above. While pounds-force or kg-force are preferred in the industry, Newton (N) is the required unit for scientific writing.

The conversion factors are as follows:
Pound-force (lbf) x 4.448 = Newton (N)
Kilogram-force (kgf) x 9.807 = Newton (N)

Maintenance of firmness testers

1. Before use each day, work the plunger in and out for 10 seconds to loosen up the springs inside the instrument.
2. Clean the tips after use to prevent clogging of the mechanism with juice.

Calibration of firmness testers

In order to calibrate firmness testers, follow these procedures:

1. Hold the firmness tester in a vertical position and place the tip onto the pan of a scale.
2. Press down slowly on the firmness tester until the scale registers a given weight, and then read the firmness tester.
3. Repeat this comparison 3 to 5 times. If you find that the instrument is properly calibrated, it is ready to use.
4. If the instrument is not in agreement with the scale, find out the magnitude and direction of the differences and proceed as follows:

10.1 Fruit firmness instruments

There are several firmness testers available; some for measuring penetration force, and others for measuring deformation forces.

1. *Penetration force testers (penetrometers)*

Penetration force testers are used for measuring firmness in some fruits such as apple.

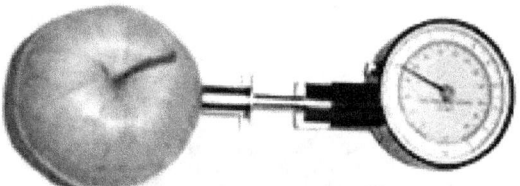

Figure 3-9: Penetrometer

There are three basic types of penetrometers available.

a. *Magness-Taylor pressure tester*: This is a device for measuring fruit firmness by reading the pressure resistance offered to compression by the fruit in pounds-force (lbf). There are two types of this tester; the original Magness-Taylor slide rule-type and spring-loaded type. The slide-rule type is reliable, but bulky and heavy.

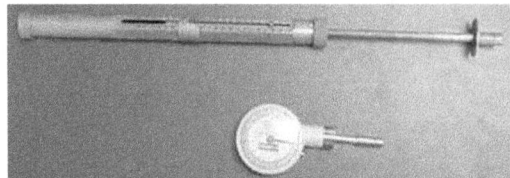

Figure 3-10: Magness-Taylor pressure tester

Procedure for using Magness-Taylor pressure tester:

⇨ Remove the plunger assembly from the barrel of the instrument and remove the bolt and washers from the end of the plunger assembly.
⇨ Pull the plunger and spring out of the metal cylinder, and then shake the washers out of the cylinder.
⇨ To make the instrument read lower, move washers from inside to outside the metal cylinder.
⇨ To make the instrument read higher, move washers from outside to inside the metal cylinder.
⇨ Reassemble and recheck for calibration.

b. *UC fruit firmness tester:* The UC firmness tester is a penetrometer that has a force sensor component housed in a frame (drill press stand). A lever fixed to the frame is used to depress the penetrometer tip (3 mm diameter) into the cherry flesh, and a firmness value is given by a gauge on the device.

Figure 3-11: UC Penetrometer

The maximum force applied to the fruit, occurring momentarily prior to collapse of the flesh beneath the penetrometer tip, is defined as the firmness value. This was the only device which caused obvious permanent fruit damage.

c. *The Effe-gi penetrometer* is lightweight hand-held probe with gauge for pounds-force. It is easy to carry with an easy to read the dial. Mounting the force gauge on a drill-press stand, as seen in the UC firmness tester, increases the potential accuracy of results. Remove the peel before compression unless the peel is the tissue of interest for firmness measurement (usually is not).

Figure 3-12: The Effe-gi penetrometer

The probes used in the instruments described above can also be mounted on computerized texture analyzers, which eliminate much operator variability. This allows not only determination of maximum force values, but also a texture profile. For example, a texture profile can show differences in the texture of chilled and no-chilled products. The most common way to measure firmness is by measuring the fruits' resistance to compression in pounds-force (lbf).

Procedures for using Effe-gi fruit penetrometer include:

⇨ Unscrew the chrome guider nut to remove the plunger assembly.
⇨ To make the instrument read lower, insert washers between the spring and the stationary brass guide.
⇨ To make the instrument read higher, insert washers between the chrome guide nut and the stationary brass guide on the plunger shaft.
⇨ Reassemble and recheck for calibration.
⇨ If the indicator needle does not stop or does not release button hold, remove the plunger assembly, and then lubricate the inside of the instrument with an aerosol lubricant.

Firmness measurements may be useful for some fruit vegetables (melons, peppers) and even root vegetables (carrots, potato), but other measurements of texture are needed for stem and leafy tissues such as asparagus or celery (force for a blade to cut or shear). For lettuce, because of the variability of the structure of the leaves, it has been difficult to develop a standard assessment of crispness.

2. *Deformation force tester*

A deformation tester is used for determining fruit firmness by measuring the deformation force for soft fruit such as tomatoes, papayas, and pears.

10.2 Measuring soluble solids content (ssc) in fruits

Sugars are the major soluble solid in fruit juice and therefore soluble solids can be used as an estimate of sugar content. Organic acids, amino acids, phenolic compounds, and soluble pectins also contribute to soluble solids. Soluble solids content (SSC) can be determined in a small sample of fruit juice using a refractometer (Figure 3-13).

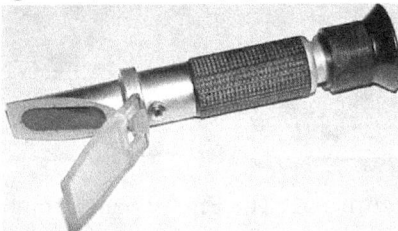

Figure 3-13: Refractometer

The refractometer measures the refractive index, which indicates how much a light beam will be slowed down when it passes through the fruit juice. The refractometer has a scale for reading the refractive index and another scale for measuring equivalent °Brix or SSC percentage which can be read directly. Digital refractometers remove potential operator error in reading values.

Figure 3-14: Digital refractometer

For small products such as cherries, strawberries and grapes, the entire fruit can be juiced. For larger products, a sample wedge should be cut from stem to blossom end and to the center of the fruit to account for variability in SSC from top to bottom and inside to outside of the fruit. A garlic press works well for small samples. Cheese cloth may be necessary to remove pulp from the juice.

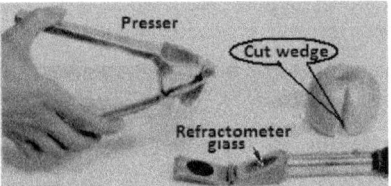

Figure 3-15: Garlic press

A wedge is cut from the commodity from stem to blossom end and to the center. The juice is extracted with a garlic press and a few drops are placed onto the glass of the refractometer.

The refractometer is closed and held up to the light for viewing through the eyepiece. The internal scale will show the SSC of the juice.

The temperature of the juice is a critical factor for accuracy because all materials expand when heated and become less dense. For a sugar solution, the change is about 0.5% sugar for every 5.6°C (10°F). Good quality refractometers have a temperature compensation capability or at least a thermometer attached to them so that the operator can make the necessary corrections. It is essential to clean the refractometer between each reading and to standardize it with distilled water (should read a refractive index of 1.3330 at 20°C (68°F) or 0% SSC).

10.3   Fruit titratable acidity

Fruit titratable acidity (TA) can be determined by titrating a known volume of fruit juice with 0.1 N NaOH (Sodium Hydroxide) to an end point of pH = 8.2 as indicated by phenolphthalein indicator or by using a pH meter. (NaOH is added to the juice until the pH changes to 8.2. The milliliters of NaOH needed are used to calculate the TA). The TA, expressed as percentage malic, citric or tartaric acid, can be calculated as follows:

$$TA = ml\ NaOH\ x\ N(NaOH)x\ \frac{Acid\ m_{eq}\ factor}{Juice\ titrated}\ ....3.1$$

Use the acid milli-equivalent factor, m_{eq} for the predominant organic acid in the commodity. The following table shows how to calculate titratable acidity, TA for 3 organic acids.

Table 3-3: Predominant organic acids use for TA calculations

Acid	Weight formula	Equivalent Wt	Acid M_{eq} factor	Commodities found
Citric	192.12	64.0	0.064	Pineapple berries citrus fruits
Malic	134.09	67.05	0.067	Apple, pear, peach, tomato
Tartaric	150.08	75.04	0.075	Grape

10.4   Maintenance of quality/quality assurance

In the fresh fruit and vegetable market, meeting a standard requires that appropriate quality-assurance methods should be in place to govern the production and postharvest handling of the product. In order to maintain these standards it is essential that an appropriate quality-assurance program is established.

In this regard the ISO 9000 series of standards is a major aid in helping producers to establish their own procedures. However, these standards are not designed for food crops in particular. Rather they are guidelines to setting and operating standards, rather than quality standards themselves. For every crop it is helpful if clear standards are defined, as this gives producers clear indications of what is needed.

Total quality management requires that quality definitions are discussed and agreed by everyone in the production chain, and that a quality manual be produced for use by the company concerned. Quality checks are required throughout the postharvest chain. These include not only checks on the quality of the product but also an assessment of the procedures and facilities at each part of the handling chain. Checks on product quality normally are undertaken whenever the product changes hands or has completed a unit operation in the chain.

Maintaining quality and extending life of harvested products

Low-temperature storage is the major tool that the post harvest operators have to maintain quality and extend life of harvested products. Low temperatures not only reduce respiration rate, but also reduce water loss through transpiration and nutritional loss. Immediately after harvest, products should be placed in a well-ventilated and shaded environment to prevent large temperature increases that could occur if products are exposed to direct sunlight.

Quality maintenance during storage and transportation

Quality of products can be maintained during storage and transportation by observing the following practices:

1. *Special considerations*

Transporter must be aware of special requirements for transporting organic product whether by highway truck, air carrier, or containerized marine and intermodal shipping. Mixed-load shipment of organic and conventional product is permitted if "Organic" labeling is prominently and clearly displayed.

In addition, there must be no risk that organic commodities will be contaminated by or come into direct contact with conventional product. Typically, carriers of bulk, raw organic product must maintain complete records of clean-out dates and products. Procedures for transport carrier cleaning or other treatments must include steps to prevent contamination from cleaners or fumigants, ripening agents, pest control agents, diesel fumes, and vehicle maintenance products.

2. *Ethylene treatment*

Another method of maintaining quality is ethylene treatment. Ethylene is a natural hormone produced by plants and is involved in many natural functions during development, including ripening. The management of ethylene may be another postharvest consideration for quality maintenance during storage and transportation. Ethylene treatments may be applied for *degreening* or to accelerate *ripening* events in fruits harvested at a mature but unripe developmental stage. External ethylene will stimulate loss of quality, reduce shelf life, increase disease, and induce specific symptoms of ethylene injury, such as the following:

⇨ Russet spotting of lettuce
⇨ Yellowing or loss of green colour (for example, in cucumber, broccoli, spinach)
⇨ Increased toughness in turnips and asparagus spears
⇨ Bitterness in carrots and parsnips
⇨ Softening, pitting, and development of off-flavour in peppers, summer squash, and watermelons
⇨ Browning and discolouration in eggplant pulp and seed
⇨ Discolouration and off-flavour in sweet potatoes
⇨ Increased ripening and softening of mature green tomatoes

3. Sanitation

Sanitation is of great importance to produce handlers, not only to protect produce against postharvest diseases, but also to protect consumers from food borne illnesses. *E. coli* 0157:H7, *Salmonella, Chryptosporidium, Hepatitis,* and *Cyclospera* are among the disease-causing organisms that can be transferred via fresh fruits and vegetables. The use of a disinfectant in wash water can help prevent both postharvest diseases and food borne illnesses. Sanitization can be achieved by adopting one of the following processes:

Chlorination: Chlorine in the form of a sodium hypochlorite solution or as a dry, powdered calcium hypochlorite can be used in hydro-cooling or wash water as a disinfectant. Some pathogens such as *Chryptosporidium*, however, are very resistant to chlorine, and even sensitive ones such as *Salmonella* and *E. coli* may be located in inaccessible sites on the plant surface. For the majority of vegetables, chlorine in wash water should be maintained in the range of 75–150 ppm (parts per million.) The antimicrobial form, hypochlorous acid, is most available in water with a neutral pH (6.5 to 7.5).

The effectiveness of chlorine concentrations are reduced by temperature, light, and interaction with soil and organic debris. The wash water should be tested periodically with a monitoring kit, indicator strips, or a swimming pool-type indicator kit. Concentrations above 200 ppm can injure some vegetables (such as leafy greens and celery) or leave undesirable off-flavours. Organic growers must use chlorine with caution, as it is classified as a restricted material.

Ozonation: This is another technology used to sanitize produce. A naturally occurring molecule, ozone is a powerful disinfectant. Ozone has long been used to sanitize drinking water, swimming pools, and industrial wastewater. Ozone not only kills whatever food borne pathogens might be present, it also destroys microbes responsible for spoilage. A basic system consists of an ozone generator, a monitor to gauge and adjust the levels of ozone being produced, and a device to dissolve the ozone gas into the water.

Hydrogenation: Hydrogen peroxide can also be used as a disinfectant. Concentrations of 0.5% or less are effective for inhibiting development of postharvest decay caused by a number of fungi. Hydrogen peroxide has a low toxicity rating and is generally recognized as having little potential for environmental damage.

10.5   Types of products contaminants

Naturally occurring contaminants: These include cyanogenic glucosides in lima beans and cassava, nitrates and nitrites in leafy vegetables, oxalates in spinach, and thioglucosides in cruciferous vegetables. Traditional processing and cooking practices usually remove the bulk of these contaminants, but there could be problems if new consumers are introduced to new crops.

Natural contaminants: Mycotoxins (fungal toxins), bacterial toxins, and heavy metals occur naturally and can be present in some crops.

Synthetic toxicants: Agricultural chemical residues, and introduced environmental pollutants such as lead, can be a problem unless production systems are controlled and monitored carefully.

Microbial contamination: Bacteria can be introduced onto fresh fruit and vegetables through the use of untreated organic fertilizers (e.g., manure), or through insufficiently treated wastewater. Inadequate hygiene standards in packing sheds and anywhere else in the food chain can also cause problems. The problem is exacerbated because fruit and vegetables often are eaten fresh. Washing fresh produce is a help, but the water used should also be clean and free of contaminants.

10.6   Cosmetic appearance

The appearance of fruit and vegetables has a major influence on perceived quality. Quality is defined in terms of shape, colour, size, uniformity, and absence of apparent defects. Thus, in the case of bruising in apples, the surface size and appearance of the bruise is more important in the fresh fruit market than the volume of the bruise. Bruises may be present, but if they cannot be seen, they do not appear to be of importance to the market. On the other hand, fruit that appears to be bruised or marked, but is not, loses value. Shape and appearance are important quality factors.

PART 2

POSTHARVEST SYSTEMS MANAGEMENT & TECHNOLOGIES

CHAPTER 4

POSTHARVEST MANAGEMENT OPTIONS

Content: Introduction to postharvest operations, machines, descriptions, utilization, packaging and management options

4.   Introduction

The inherent quality of produce cannot be improved after harvest, only maintained for the expected window of time (shelf life). Part of what makes for successful postharvest handling is an accurate knowledge of what windows of opportunity are available under specific conditions of production, season, method of handling, and distance to market.

In organic production, farmers harvest and market their produce at or near peak ripeness more commonly than in many conventional systems. However, organic production often includes more specialty varieties whose shelf lives and shipping traits are reduced or even inherently poor. In general, the level of susceptibility of these products to handling damage is greatly underestimated, usually because the effects of mishandling do not appear until sometime after the damage occurred. Poor handling and storage can easily result in a total crop loss anywhere.

Most fruit and vegetable crops begin to deteriorate as soon as they are harvested, and most are particularly prone to handling damage at all times. Not all losses are due to poor handling, but handling damage is known to accelerate other types of deterioration, particularly the development of molds and rots. As a general approach, postharvest practices can help maintain product quality.

4.1 Postharvest practices

Post harvest technologists attempt to maintain quality by slowing deterioration to improve storage and shelf life, and thus ensuring consumers have high quality fruits and vegetables to purchase. They seek to control the handling, transport, and storage conditions to ensure optimal quality and as such need to take account of the living nature of horticultural products, and in particular their susceptibility to physical, pathological, and physiological deterioration.

Evidently, postharvest practices differ among crops and some crops may or may not require all the stages required to enhance the quality of such product. An abuse of physical factors involved in handling product can lead to increased postharvest loss of quality. Post harvest operations involved in product handling and processing are described below.

4.1.1 Transfer from field bin to reception

Transfer from field bin

Fruit and vegetables must be transferred from the field to arrive in a state that is acceptable to the consumer. Crops such as apples are collected into larger containers (field bins) which are transported out of the orchard. The transfer into the field bin is a serious potential cause of damage, unless the pickers are well trained. Fruit-on-fruit impact and impacts against the sides and base of the bin are potential sources of severe bruising. The fruits were transported to packing shed over rough farm tracks and other roads to the grading shed or market. This movement causes potential damage to fruit as there is no packaging to protect them.

Figure 4-1: Harvested fruits ready to be transported in field bins

Unloading from field bin or trailer

This operation needs to be achieved with minimum of damage. Floating out fruits into channels using water works well for fruit with a density less than that of water apples and Pears can be removed in the same way, providing a suitable chemical is added to the water to increase its density. If water cannot be used, then options include tipping the bin over slowly or using side openings on the bin. Submerging apples in the field bin in water results in little damage.

Figure 4-2: Unloading fruits from field bin

Reception

At the reception, documentations on where and who the product is from, the harvest date and the delivery date to the shed, as well as full details about the crop and pesticides used during its growth of the crop arriving at a grading shed is recorded. This allows the grower to receive appropriate credit for the product and ensures that quality guidelines have been adhered to prior to arrival of the product at the shed. Reliable documentation requires careful inventory control, including clear procedures for recording all shipments and marking the field bins.

Figure 4-3: Receiving fruits at juice plant for immediate processing or storage

4.1.2 Product cooling at reception

Precooling

If the product cannot be dealt with immediately when it arrives at the packing shed, it is essential to minimize the deterioration of product. For most crops this involves reducing the temperature. Highly perishable leafy vegetable crops can be cooled rapidly in the field using mobile vacuum coolers

Figure 4-4: Hydro precooling with water spray

Drenching

Drenching is done to reduce physiological disorders or infestations in harvested products. This can be done using an overhead spray system that washes through the field bin, or dipping the bin in a bath, which may be outside the grading shed.

Figure 4-5: Single-bin drenching with a postharvest fungicide

Washing

Field dirt must be removed before sorting. Providing fruit is not seriously soiled, the water dump followed by a series of rotating brushes will remove dirt, without causing damage to the fruit. However, there is some concern that brushing can increase water loss from some products such as citrus. Also, the surface of some fruits (especially pears) is easily damaged by brushing or rubbing.

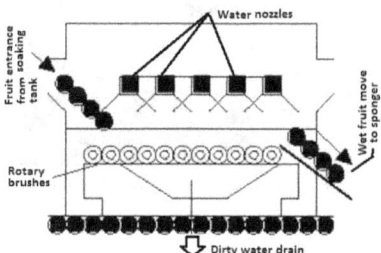

Figure 4-6: Typical produce washing machine

Waste water removal

After washing, surplus water should be removed. This usually is achieved as the fruit passes over the brushes or absorbent foam rollers but may require forced warm air from fans.

4.1.3 Sorting & grading

Presorting

Presorting is used to remove fruit that clearly does not meet specified quality standards. This process helps to maximize throughput in the rest of the grading system.

Figure 4-7: Presorting

Machine vision colour sorters remove poor coloured fruits, and *presizers* remove excessively small or large fruit, before manual grading of fruit according to quality standards.

Sorting/grading

During sorting, fruit are graded (sorted) into categories. In many cases there are only two or three grades (e.g., export, local market, or juicing). Although computer vision systems are making significant advances, grading still is best achieved in most cases by human inspection.

Figure 4-8: Apple sorter

Sorters require suitable facilities to enable them to see defects clearly. Fruit must be rotated so that the sorters can see the entire fruit surface. Sorters will perform best if they work under good lighting conditions and in comfortable working positions.

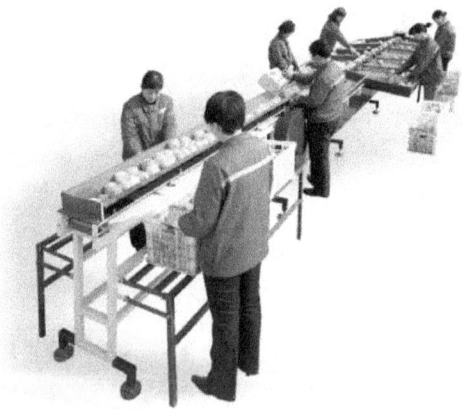

Figure 4-9: Fruit grading

Singulating

If fruit are to be sorted according to weight, the singulator separates fruit into pockets or cups so that each fruit can be weighed independently. The fruit can then be separated into appropriate sizes by sorting according to the weight recorded for the cup. Various devices including transfer wheels and expanding belts can be used for singulation.

Figure 4-10: Sinclair multi-head singulator

Size band sorting

Size uniformity and conformity is particularly desirable for the purposes of fruit packaging and display. Some fruits and vegetable have a consistent shape, thus they can be conveniently weighed and sorted. The result is a product that is consistent in volume and shape and packs easily. Electronic or mechanical methods can be used for weighing; shape sorters can be classified as mechanical or visual.

Figure 4-11: Band sorter

Colour & size sensor: Colour & size sensor can be incorporated into any commercial sorting line which uses a cup-conveyor or roller-conveyor for transporting the fruits. Each colour & size sensor incorporates three sets of photodiodes which view each fruit from three directions. The fruit size is determined by the average of the signals from the 3 photodiodes while most of the fruit's surface is included in the colour determination. As the fruit passes underneath the photodiodes the computer acquires about 1,000 readings from each sensor. The colour and size values are averages of these readings. A simple air nozzle separates the fruit stream into two colour classes. In this example, Star-king apples are sorted into red and pink grades.

Figure 4-12: Colour and size sensor

Electronic weighing

Fruits can be weighed with either mechanical weighing scales or electronic weight sorters. In mechanical weighing systems, a quantity of fruit is arranged in the weighing tray and the resistance to weighing spring monitored by the needle and read on the graduation is recorded.

Figure 4-13: Weighing scales

In the electronic system all fruit are weighed at one point. Data is fed to a computer, which selects and preprograms the drop point. Fruit may rest on several fingers (rather than a single cup, as in Figure 4-14), which are programmed to release the fruit at the required point, and the fruit rolls off sideways.

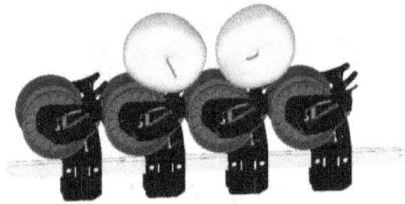

Figure 4-14: Electronic weight sorter

Mechanical weighing

In a mechanical system, each fruit is "weighed" as it passes each exit point. A mechanical lever releases the fruit if the weight exceeds the preset trigger level at that point. Drop points therefore must be arranged along the line in increasing weight order, which can cause logistical problems. Mechanical systems are usually less costly and can be maintained and repaired by a trained mechanic.

Figure 4-15: Circular weight grader with shallow bins

Size sorting

Some crops (e.g., tomatoes, carrots) produce inconsistent results if sorted by weight. Better results are obtained if they are sorted according to shape, using expanding screens or tapers. Fruit moves through the sorter and passes over or rolls along a continuously enlarging exit orifice until it is able to fall through into the appropriate container, such as in Figure 4-16.

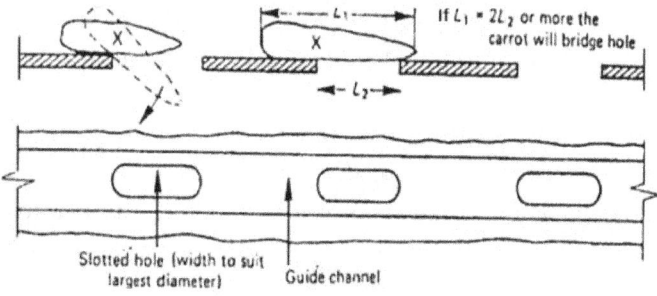

Figure 4-16: Fruit length grading mechanisms

In the field, crops also can be sized by eye, or by using a set of rings or plastic cards of predetermined size.

Image analysis sorting

Increasingly, image analysis is used to size fruit and vegetables. Cameras take a view of the product, and according to the algorithm the fruit is sorted according to cross-sectional area or according to some other shape factor (e.g., length).

Waxing

Fruit waxing of is done to add value and improve the appearance of the fruit to the customer. Coatings also may reduce water loss and affect long-term colour changes. Waxing is achieved by passing the fruit under wax-solution sprayers, combined with rotation on expanded plastic foam rollers already soaked by the spray.

4.1.4 Packaging & labeling

Packaging

Once sorted into grades and size, fruit are delivered by conveyor to a packing area. Here fruit may be held in rotating final size bins until they can be packed by hand into cartons. Alternatively, if automatic tray fillers are used, fruit fall into a paper pulp tray directly. Fruit and vegetables also may be wrapped individually in foam liners or films before, or instead of, packing them in a rigid container.

Figure 4-17: Pressure-pack tray system

Packaging fulfils several functions including containment, facilitating transportation, protection of fruit from further damage, protection of the environment from contents of package (for example, if the contents are dirty), marketing, product advertising, and stock control.

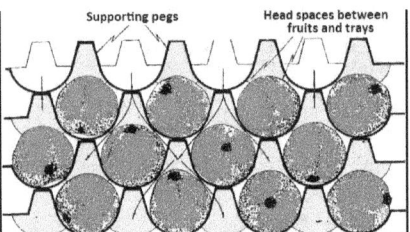

Figure 4-18: Peg-type tray system

Labeling

On the international market, some fresh fruits require considerable detail. This may include, for example, harvest date, packing date, and name of grower and packer, as well as details about the cultivar and grade. This information can be critical if quality management problems arise in the retail market.

Libelling packages helps handlers to keep track of the produce as it moves through the postharvest system, and assists wholesalers and retailers in using proper practices. Labels can be pre-printed on fibreboard boxes, or glued, stamped or stencilled on to containers. Brand labelling packages can aid in product packer, and producer advertisement. Some shippers also provide brochures detailing storage methods or recipes for consumers.

Labelling of consumer packages is mandatory under national food and drugs administration (FDA) regulations. Labels must contain the name of the product, net weight, and name and address of the producer, packer or distributor.

Shipping labels contain some or all of the following information:

a. Brand name.
b. Common name of the product.
c. Net weight, count and/or volume.
d. Name and address of packer or shipper.
e. Country or region of origin.
f. Size and grade.
g. Recommended storage temperature.
h. Special handling instructions.
i. Names of approved waxes and/or pesticides used on the product and
j. Probable shelve life.

4.1.5 Quality control checks

Quality-control checks must be carried out before the fruit leaves the packing shed. Although a check should be made once the fruit has been packed, good quality management requires that the process be continuous and cover all parts of the chain. Thus problems can be identified early and corrective action taken.

4.2 Postharvest management options

Foods (fruits and vegetable) in postharvest operation should be well managed or handled to preserve as well as maintain its harvest quality. Some of the postharvest management options available are described below.

Precooling

Pre-cooling is the first step in good temperature management. The *field heat* of a freshly harvested crop is usually high, and should be removed as quickly as possible before shipping, processing, or storage. Previously cooled fruit will also have warmed up during sorting and will need re-cooling. For rapid cooling, it is necessary to place fruit in a precooler for fast cooling to room temperature before transferring to the normal cold store/room.

Figure 4-19: Inside view of a precooler

Refrigerated trucks are not designed to cool fresh commodities but only maintain the temperature of pre-cooled produce. Precoolers are designed to extract heat as quickly as possible and desirable by using forced air flow through the load to be cooled.

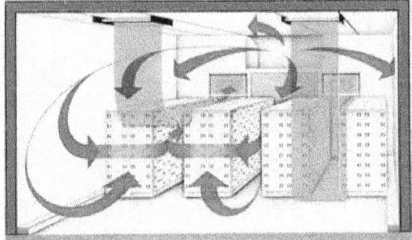

Figure 4-20: Air-flow through precooler

Methods of precooling produce

There are seven principal methods of pre-cooling fresh produce:

1. *Room cooling*: This is a widely used precooling method involving the placing of produce in boxes (wooden, fiberboard or plastic), bulk containers or various other packages in an insulated room equipped with refrigeration unit, where they are exposed to cold air. It is used for produce sensitive to free moisture or surface moisture. Because this type of cooling is slow, room cooling is only appropriate for very small amounts of produce or produce that does not deteriorate rapidly.

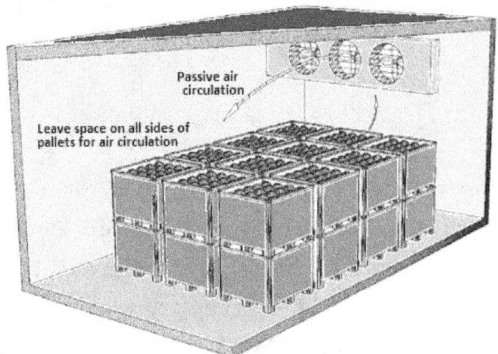

Figure 4-21: Room cooling

Typically the cold air is discharged into the room near the ceiling, and sweeps past the produce containers to return to the heat exchangers. The cooled air is generally supplied by forced or induced draft coolers, consisting of framed, closely spaced and finned evaporator coils fitted with fans to circulate the air over the coils.

2. *Forced-air cooling:* Forced air cooling was developed to accommodate products requiring relatively rapid removal of field heat immediately after harvest. Forced air or pressure cooling is a modification of room cooling and is accomplished by exposing packages of produce to higher air pressure on one side than on the other. This technique involves definite stacking patterns and the baffling of stacks so that the cooling air is forced through (rather than around) the individual containers. For successful forced air cooling operations, it is required that containers with vent holes be placed in the direction of the moving air and packaging materials that would interfere with free movement of air through the containers should be minimized.

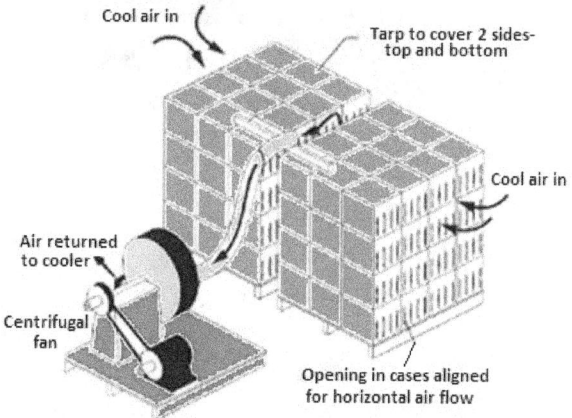

Figure 4-22: Forced horizontal air flow

Air can be channeled to flow either horizontally or vertically. In a horizontal flow system, the air is forced to flow horizontally from one side of the pallet load to the other through holes in the sides of the pallet bin or containers.

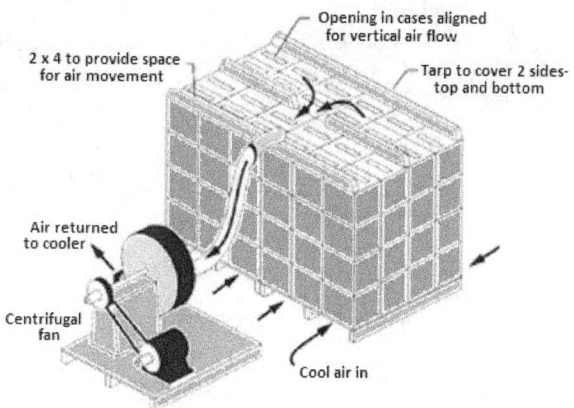

Figure 4-23: Forced vertical air flow

In a vertical flow system, the air is forced to flow vertically from the bottom to the top of the pallet through holes in the bottom of the pallet, and containers if used, then out the top.

The following are forced-air cooling alternatives:

Cold wall: A permanent false wall or air plenum contains an exhaust fan that draws air from the room and directs it over the cooling surface. The wall is at the same end of the cold room as the cooling surface. The wall is built with a damper system that only opens when containers with openings are placed in front of it. The fan pulls cold room air through the container and contents, cooling the produce.

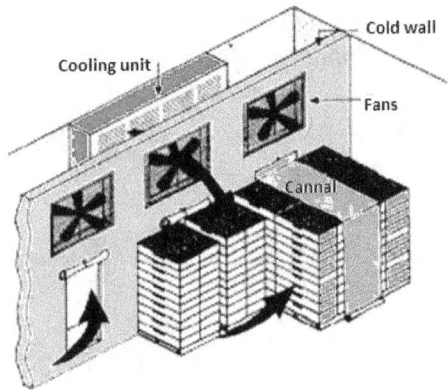

Figure 4-24: Cold wall

Forced-air tunnel: An exhaust fan is placed at the end of the aisle of two rows of containers or bins on pallets. The aisle top and ends are covered with plastic or canvas, creating a tunnel. An exhaust fan draws cool room air through the container vents and top . The exhaust fan may be portable, creating a single forced-air tunnel where needed, or it may be part of a stationary wall adjacent to the cooling surface, with several fans that create several tunnels.

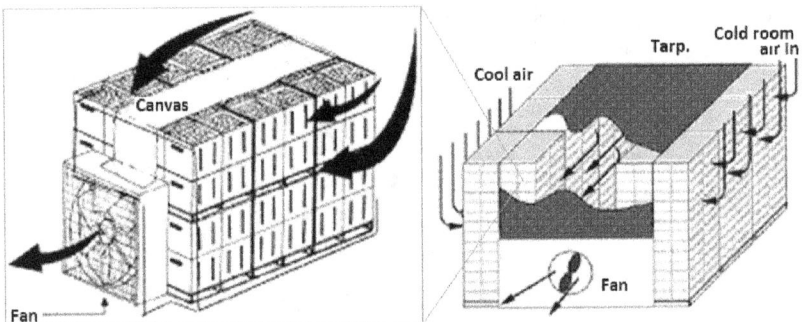

Figure 4-25 Forced-air tunnel

Serpentine cooling: A serpentine system is designed for bulk bin cooling. It is a modification of the cold-wall method. Bulk bins have vented bottoms with or without side ventilation. Bins are stacked several high and several deep with the fork lift openings against the cold wall. Every other forklift opening—sealed with canvas—in the stack matches a cold wall opening. The alternate unsealed forklift opening allows cold air to circulate through the produce. Cold room air is drawn through the produce via the alternate unsealed openings in the stack and the top of the bin.

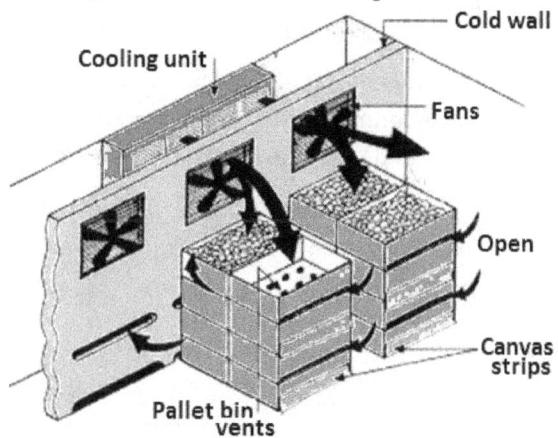

Figure 4-26: Serpentine cooling

3. *Hydro-cooling:* Hydrocooling is the utilization of chilled or cold water for lowering the temperature of a product in bulk or smaller containers before further packing. Hydrocooling is achieved by flooding, spraying, or immersing the product in/with chilled water. Hydrocooling methods differ in their cooling rates and overall process efficiencies. There are several different hydrocooler designs in operation commercially. Differences between the individual techniques are evident by the method of cooling and by the way that produce is moved or placed in the cooler.

Figure 4-27: Hydro-cooling with water spray

Various types of hydrocooler available include the conventional (flood) type, immersion type, and batch type.

Conventional (flood) type: The flood type hydrocooler cools the packaged product by flooding as it is conveyed through a cooling tunnel. A frequent complaint about both conventional and batch type hydrocoolers is that cooling by these techniques is not uniform and hence may leave `hot spots' throughout the load. For hydrocooling to be effective, contact between the water and the product surface must be uniform.

Figure 4-28-: Conventional system

Immersion type: The bulk or immersion type cooler uses a combination of immersion and flood cooling. Loose produce is immersed in cold water, and remains immersed until an inclined conveyor gradually lifts the products out of the water and moves it through an overhead shower.

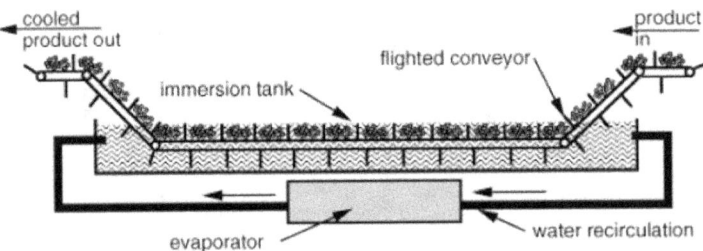

Figure 4-29: Cut-away side view of a continuous-flow immersion hydrocooler

Shower coolers: Shower cooler distribute water using a perforated metal pan that is flooded with cold water from the refrigeration evaporator. Shower type coolers can be built with a moving conveyor for continuous flow operation or they can be operated in a batch mode.

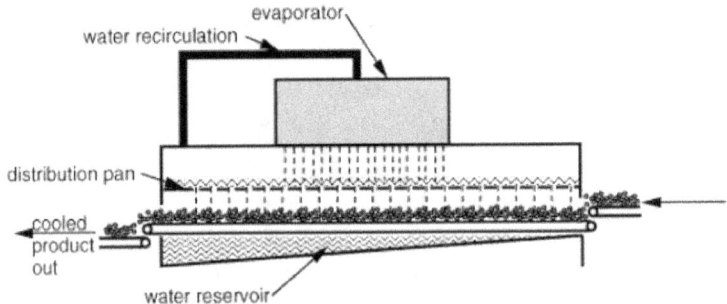

Figure 4-30: Continuous-flow shower-type hydrocooler

Batch system: With the batch system, chilled water is sprayed over the product for a certain length of time, depending on the season and the incoming product temperature. These hydrocoolers have a smaller capacity than conventional hydrocoolers and are therefore less expensive.

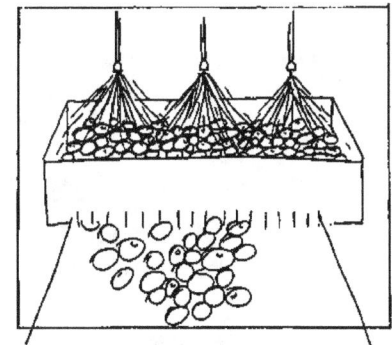

Figure 4-31: Batch hydrocooling system

HydroJet precooling: The hydroJet technology is specially developed to keep the quality of your product by cooling your product quickly without dehydration. Besides this you can fill your storage in the same speed as your harvest and start cooling directly.

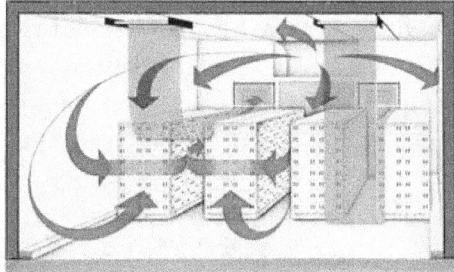

Figure 4-32: Hydrojet precooler

4. *Ice cooling:* In ice cooling, crushed or fine granular ice is used to cool the produce. The ice is either packed around produce in cartons or sacks (a process called top icing), or it is made into slurry (liquid icing) with water and injected into waxed cartons packed with produce. The ice then fills the voids around the produce.

Figure 4-33: Crushed or flaked ice for package icing

The use of ice to cool produce provides a high relative humidity environment around the product. Before the advent of comparatively modern precooling techniques, contact or package icing was used extensively for precooling produce and maintaining temperature during transit. There are a variety of different methods (Figure 4-34) in which ice are applied to the produce so as to achieve the desired cooling effect.

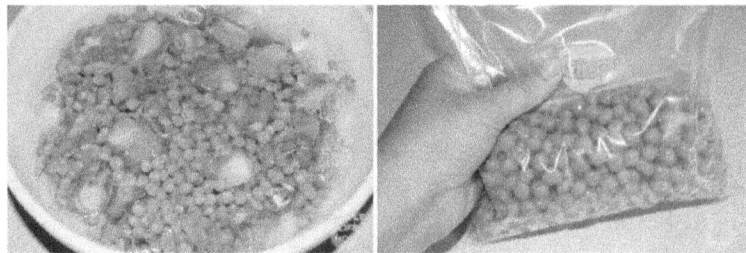

Figure 4-33: Ice coolers

Package ice can be used only with water tolerant, non-chilling sensitive products (such as: carrots, sweet corn, cantaloupes, escarole, lettuce, spinach, radishes, broccoli, green onions). and with water tolerant packages (waxed fiberboard, plastic or wood).

5. *Vacuum cooling:* Vacuum cooling is achieved by the evaporation of moisture from the product. Produce is enclosed in a chamber in which a vacuum is created. As the vacuum pressure increases, water within the plant evaporates and removes heat from the tissues. This system works best for leafy crops, such as lettuce, which have a high surface-to-volume ratio. The evaporation is encouraged and made more efficient by reducing the pressure to the point where boiling of water takes place at a low temperature.

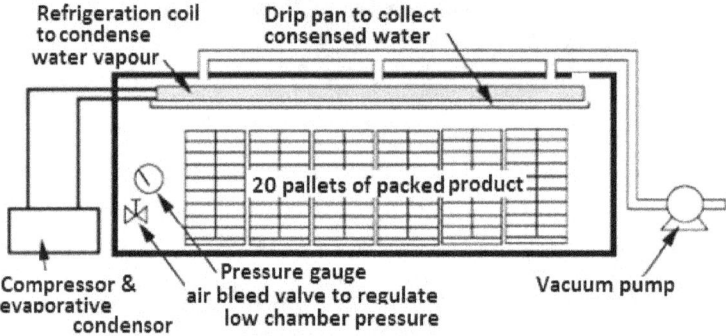

Figure 4-34: Vacuum precooler

Figure 4-34: Inside view of a vacuum precooler

Vacuum cooling process takes place in three steps:

a. *Pressure reduction*: At atmospheric pressure (1013 mbar), the boiling temperature of water is 100 °C. The boiling point of water changes as a function of saturation pressure therefore at 23.37 mbar the temperature at which water will boil is 20 °C and at 6.09 mbar, it will boil at 0 °C.

b. *Reduction in sensible heat*: To change from the liquid to vapour state, the latent heat of vaporization must be provided by the surrounding medium, so that the sensible heat of the product is reduced.

c. *Water drain*: The water vapour given off by the product must be removed.

6. *Cryogenic cooling:* In cryogenic cooling, the produce is cooled by conveying it through a tunnel in which liquid nitrogen or solid CO_2 evaporates. The use of the latent heat of evaporation of liquid nitrogen or solid CO_2 (dry ice) can produce `boiling' temperatures of -196 and -78 °C, respectively. This is the basis of cryogenic precooling. At these temperatures, the produce will freeze and thus be ruined as a fresh market product.

Figure 4-35: Cryogenic fruit cooler

7. *Evaporative cooling:* Evaporative cooling is an inexpensive and effective method of lowering produce temperature. It is most effective in areas where humidity is low. Dry air is drawn through moist padding or a fine mist of water, then through vented containers of produce. As water changes from liquid to vapour, it absorbs heat from the air, thereby lowering the produce temperature.

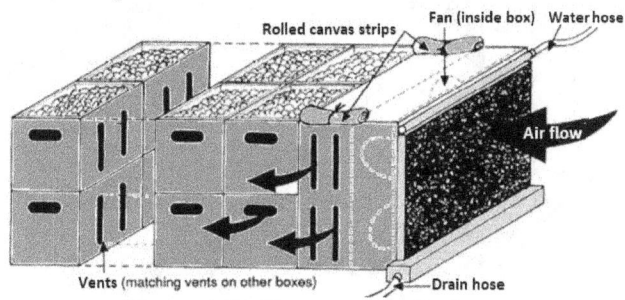

Figure 4-36: Evaporative cooling

This method has a low energy cost but cooling efficiency is limited by the capacity of air to absorb humidity. As a result, it is only useful in areas of very low relative humidity.

The incoming air should be less than 65 percent relative humidity for effective evaporative cooling. It will only reduce temperature, 10-15°F. This method would be suitable for warm-season crops requiring warmer storage temperatures (45-55°F), such as tomatoes, peppers, cucumbers or eggplant.

Cool/refrigerated storage

Once at store temperature, fruit and vegetables can be transferred to a larger cool storage. Its function is to maintain rather than reduce the temperature of the product. A refrigerated room is a relatively airtight and thermally insulated building. The refrigeration equipment should have an external escape outlet to release externally the heat generated by the product. Refrigeration capacity of the equipment should be adequate to extract the heat generated by crops with a high respiration rate. It is also important to precisely control temperature and relative humidity conditions inside the refrigerated storage environment.

Figure 4-37: Cold storage

Refrigerated rooms can be made with concrete, metal, wood, or other materials. All external surfaces should be thermally insulated, including the floor and ceilings. Type and thickness of insulation material depends on building characteristics, produce to be stored and the difference in temperature required between external and internal conditions.

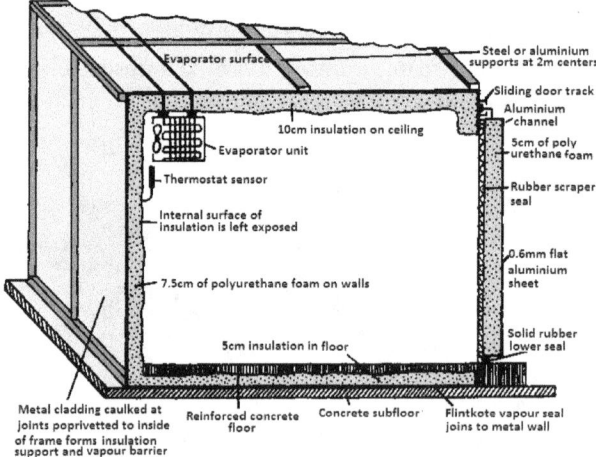

Figure 4-37: Cold storage features

Mechanical refrigeration has two main components: the evaporator, inside the storage area and the condenser which is outside connected by tubing filled with refrigerant. Normally, both elements are finned coils made of high thermal conductivity materials and integrated to a fan. This facilitates heat exchange.

Figure 53: Inside a refrigerated storage.

Product packaging: Loading and transportation

Finally, product is loaded onto appropriate transportation by forklift truck and transported to the eventual market destination. This may be days or months after packing. Although the product usually might have been packaged, it is still vulnerable to damage if subjected to rough ride over a considerable distance.

Figure 4-38: Loading final packaged product with forklift

Over long distances, refrigerated vehicles are essential. Maintaining optimum temperature in refrigerated transportation can:

- Reduce respiration
- Slow down moisture loss and wilting
- Prevent spoilage due to bacteria, fungi and yeast
- Retard undesirable growth such as sprouting

CHAPTER 5

POSTHARVEST SYSTEMS MANAGEMENT IN THE TROPICS

Content: Introduction to postharvest changes in fruits and vegetables, crop losses in the tropics, postharvest challenges in the tropics, challenges to postharvest development and solutions

5.   Introduction

Developing markets for tropical crops that are unknown in developed countries offers tropical countries a great opportunity to take advantage of the current desire for a wide variety of healthy foods. These countries have potentials to produce a wide range of food products that are capable of supplying both their domestic and overseas market needs. For example, such crops include baobab, wild mango, African fan palm, African ebony, breadfruit, star fruit, jackfruit, and sour-sop etc.

5.1   Postharvest changes in fruits and vegetables

Changes that occur in the harvested, unprocessed vegetable include: water loss, conversion to starch and sugar and vice versa, flavour change, colour changes, sprouting, rooting, softening and decay. Some changes results in quality deterioration and others improve quality in those vegetables that complete ripening after harvest.

The following factors influences changes in fruit and vegetable products after harvest

a. Kind of crop
b. Air temperature and circulation
c. Oxygen and carbon dioxide contents

    d.   Relative humility of the atmosphere

    e.   Disease and micro organisms.

5.2   Crop losses in the tropics

All fresh horticultural crops are high in water content and are subjected to desiccation (wilting, shriveling) and to mechanical injury. Various authorities have estimated that 20-30 percent of fresh horticultural produce is lost after harvest and these losses can assume considerable economic and social importance. That is why; these perishable commodities need very careful handling at every stage so that deterioration of produce is restricted as much as possible during the period between harvest and consumption.

The deterioration of fruit after it has left the tree depends on one or some of the following factors:

1.   Growth and activities of micro organisms.
2.   Activities of natural food enzymes
3.   Insects, pesticides and rodents
4.   Temperature; both heat and cold
5.   Moisture and dryness
6.   Air and in particular oxygen
7.   Light and
8.   Time of harvest.

Estimates of crop losses in the tropics vary from country to country. Although grain losses are put at around 25% in less-developed countries, wastage of fruit and vegetable crops has been estimated to be around 50% of production. This is a huge loss and represents an enormous waste of resources and opportunity. In contrast, losses in developed countries are generally around 10% to 25%.

There can be many reasons for losses. The existence of and access to a suitable market outlet at an economic price, knowledge of postharvest principles, available storage facilities, availability of labour, good handling systems, and good product management are all essential components of successful utilization of fruit and vegetables, and only if all these factors are present can losses be reduced significantly. Good postharvest practices can reduce losses substantially, but implementation requires good education, good systems, good management, and the availability of resources.

5.3 Postharvest challenges in the tropics

1. Temperature

Temperature is one of the major factors in quality loss after harvest. It is the most important factor in preserving the freshness and quality of harvested commodities. It is essential to cool fruit and vegetables as quickly as possible to the lowest temperature they can sustain without chilling injury. Unfortunately, many tropical products begin to experience this physiological disorder at around 10 to 13°C, and so this sets a lower limit to the storage temperature.

There has been some evidence to suggest that in some fruits, chilling injury can be reduced by an initial hot-water treatment (a few minutes in water at around 45°C). This may enable storage of fruit at lower temperatures. For some products, particularly local cultivars of tropical fruit and vegetables, storage and handling requirements are unknown.

Some vegetable crops (mainly root crops) have appreciable periods of dormancy before the normal aging process continues. Yams, potatoes, beets, carrots, and onions can be stored for lengthy periods without cool storage, although the extent depends on the cultivar. Some are best left in the soil, where temperatures are cooler, provided the soil is not too wet and the crops are protected from attack by animals.

Figure 5-1: Field storage of onions in heaps covered with straw

Onions and garlic can be stored hung up using twine in shaded areas. Other vegetables, such as peas and beans, can be dried. Leafy vegetables, and other vegetables deteriorate rapidly after harvest, and preservation processes such as sun drying or domestic processing using sugar, brine, or vinegar are effective. Fermentation processes also can be used to preserve cabbage, breadfruit, radishes, cassava roots, and green bananas.

Figure 5-2: Storing garlic hung in shelters with natural ventilation

Options for reducing temperature without conventional refrigeration include:

a. *Adequate ventilation:* The storage facility should be well insulated and vents should be located at ground level. Vents can be opened at night, and fans can be used to pull cool air through the storeroom. Storage structures can be cooled using night air if the difference in day and night temperature is relatively large. Suitable for crops with long storage lives, these should be built where the night-time air temperature is low during the storage period, and where the best breeze can be found.

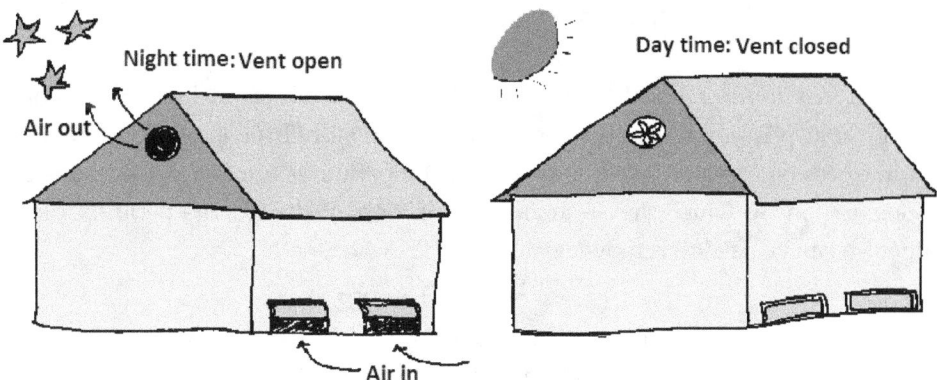

Figure 5-3: Store house with night air ventilation

The building should be orientated to make use of the breeze as much as possible. If nighttime temperatures are very low, hot-air in-flow during the day can be reduced with louvers or screens. The roof and walls should be well insulated

Grass thatch materials also can be wetted to produce evaporative cooling effect. Alternatively, white paint on exterior surfaces reduces absorption of heat. Ventilation between walls and roof and under the floor also is needed. Double-skinned walls also are better than single-layer walls.

Figure 5-4: Straw packinghouse with thatched roof

Root crops can be stored on the ground in simple clamps, consisting of a long triangular pile of the product 1 to 2 m wide, covered by straw and soil layers, and with adequate provision for water run-off. Ventilation also is provided by making a triangular open channel at the base of the pile in the center and running the length of the pile. This central duct may need vertical chimney vents at regular intervals.

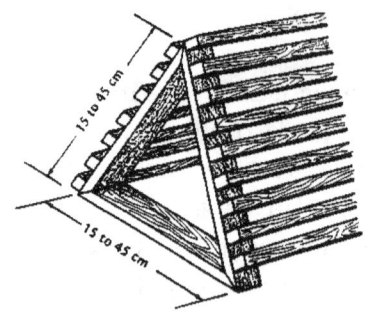

Figure 5-5: Triangular wooden duct

b. *Storage in local features*: Local features for storage of products include store room, caves, underground pits etc. Fruit and vegetables can be stored for longer periods if they are kept in stores on elevated ground, where air temperatures are lower. Caves are also used in some regions in which they remain cool throughout the storage period. Cold running water also can be used to remove heat.

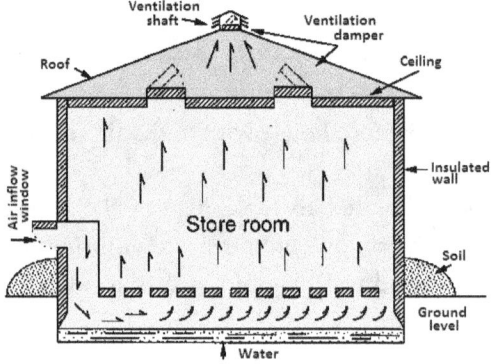

Figure 5-6: cross-sectional view of a storehouse for fruits

Illustrated in Figure 5-6 is a cross-sectional view of a storehouse for fruits. This system was officially approved as the standard model for farm-level storehouses by the Ministry of Construction (Korea) in 1983. Note that air inlets are at the base of the building, and the floor is perforated, allowing free movement of air. The entire building is set below ground level taking advantage of the cooling properties of soil.

c. *Postharvest handling procedures*: Keeping crops cool can be achieved by very simple means. If crops are harvested just before the sunrise they will be at their coolest. Once crops were harvested, they should be kept out of the sun. Thus, in the field harvested products should be placed in the shade at all times. They should be covered during daytime transportation, and trucks should not be parked in the direct sun. In the marketplace, fruit and vegetables also should be shaded. Although obvious, these practices often are not followed. Products also require ventilation, as they generate heat internally, and unless this is removed continuously, the temperature will rise and deterioration will be accelerated.

2. *Handling, packaging and transportation*

Good packaging and careful handling increase the storage life of most crops and reduce damage. The same issues apply to tropical products as any other fruit or vegetable. However, in less-developed countries problems can be exacerbated by ignorance and cost-cutting. Roads can be very rough and delays common. Distances can be great and packaging minimal.

Figure 5-7: Poor road conditions for transportation

In many markets it is considered uneconomical or impractical to use packaging, and so a high level of damage is accepted. Packaging can be produced from local biodegradable materials.

Options include bags, sacks, and woven baskets, but these offer little protection for the product. Wood boxes provide some protection but may not be acceptable on international markets. Cardboard or plastic containers also may be available. It also is possible to use secondhand containers if these are available locally.

3. *Grading systems*

In less-developed countries the majority of handling is likely to be done by hand, with limited options for mechanized systems. Machinery for automatic sizing and weighing may be appropriate, but because humans are generally better at sorting fresh produce for quality, low-cost labour can be an advantage.

4. *Storage systems*

Modern and traditional storage systems have a continuous role to play in the tropics. Modified designs of traditional structures can greatly reduce losses due to vermin and other pests. For many fruits and vegetables, drying added to their storage life and storage can be based on designs developed for cereals.

Selling season and storage life of crops that will not be transported and marketed fresh immediately after harvest can be extended into the dry months by growing crops, vegetables and fruits suited for long-term storage. The challenge is in keeping the quality high by creating and maintaining the correct storage environment. The two structural options for storage of these crops are coolers and root cellars.

Coolers used for root crop storage will require water added to the air and regular monitoring of the humidity level. Some growers have used concrete basements of houses, closed off from heat and with ventilation to let in cold dry air, as *root cellars*. Another idea is to bury a big piece of culvert under a hillside. Whatever the method, only "perfect" produce is suitable for long-term storage, so careful inspection is critical. Any damaged produce is going to spoil and induce spoilage in the rest of the crop.

5. *Waste disposal systems*

Interchangeable containers offer a convenient way of dealing with waste. The containers are placed at strategic locations and collected by truck when full, or at set times, with an empty container delivered at the same time. Local domestic systems may not be designed to cope with the volume involved, and so special arrangements are needed.

6. *Water and sewerage*

Requirements for water and sewerage in markets are higher than in other buildings due to the requirements for cleaning. For a 10,000 m^2 market, a daily demand of 75,000 liters of water is likely. This includes some produce washing and a 10,000 liters allowance for cool storage. Sanitary requirements should be around 1 unit per 15 to 25 persons.

5.5   Challenges to postharvest technology development

Critical among challenges of postharvest technology development are cultural believes, social issues etc. each of these challenges are further expatiated below.

Social issues/challenges

There have been renewed concerns in both developed and developing countries about the introduction of new technology into many development programs. Too many socio-cultural and techno-economic factors can affect the transfer and adoption of such new technologies in the developing world. It is generally agreed that those technologies most likely to be adopted by developing countries are those that are appropriate and acceptable. That is, in addition to operational and technical suitability, the technology must be profitable, affordable, available, and sensitive towards cultural, social, and environmental issues.

Culture and religion

Culture refers to a set of shared values, attitudes, and behaviours that characterize and guides a group of people. As a rule, less-developed countries tend to have more rigid social structures, stronger religious influences, very distinct gender roles, and a high diversity of languages. Certain practical implementation issues also may affect the development, testing, evaluation, and application of quality-control practices in the fresh-produce industry of developing countries.

Similarly, religious beliefs cannot be ignored in determining the acceptability of new or improved technologies. The cultural diversity of the consumer and supplier may be at variance; this makes the training of staff and the adherence to quality standards particularly difficult.

Labour

Generally, in the developing economy, skilled labour is scarce and unskilled labour abundant. Education levels are low in many less-developed countries, although some have achieved nearly universal literacy. Training for quality control can be hampered by a lack of basic education and the inability to understand the need for quality standards. However, much grading of fruit and vegetable produce relies on the human eye; with abundant labour, it should be easier to select, train, and employ people in less-developed countries.

Infrastructure

The efficiency of transportation systems, high frequency of equipment breakdown due in part to unsuitable operating conditions, inadequate repair and maintenance facilities, and power supplies has a major impact on quality-assurance procedures. These had affected the manufacture of some or all equipment in the developing country. As much as possible, In addition to providing for the participation of the population, local manufacture should be promoted and integrated into social structure through the growth of industries associated with raw materials supply and marketing.

Marketing constrains

Marketing constraints include a lack of information concerning quality requirements and packaging, poor quality of produce, and inadequate quality standards. The managerial implication of a lack of infrastructure is that the export enterprise must take on the burden of providing or improving its own infrastructure (e.g., establishing information networks, purchasing transportation systems).

Institutional factors

The political fluidity and bureaucracies of some less-developed countries has weakened the institutions of government. They are frequently overstaffed, underpaid, and in some instances under qualified. The implications of this for postharvest quality are inefficient, slow, and costly government services that have the task of approving export shipments. Centralization of governmental decision making may prevent an export company from moving quickly in response to market demand. There also may be a lack of flexibility in some educational institutions that could or should be key players in training staff.

Management structures

Both less-developed and developed countries need to ensure that there is a product quality management component in agricultural-development programs. Individual farmers are unlikely to develop the grading, storage, and distribution systems for their produce that will be demanded by the international market. Cooperative systems may be able to make progress towards introducing quality-assurance systems, but in general many less-developed country farmers have demonstrated an innate resistance to participate in co-operatives.

5.6   Solutions to challenges of postharvest technology development

In order to meet existing international standards so as to compete favorably in the sophisticated 'world market' of fresh fruits and vegetables, the following must be ensured.

Awareness

Managers should be kept fully aware of the happenings and developments in international trading agreements and regulations. For example there may be considerable advantages in marketing by gaining ISO 9000 certification for quality-management procedures.

Development effective management of technological processes

Numerous studies on new technologies to a society show how deeply interdependent the technical, economic, political, social, and cultural factors involved in the process of technical change. Thus, the understanding and effective management of the process of technological change requires an approach involving the expertise and analytical skills of several disciplines from the engineering to the social sciences. It is also important to emphasize that less-developed countries should not be treated as a homogenous group, because particular strengths and weaknesses vary widely between countries.

Development of new management opportunities

The development and implementation of relevant local contents policy and quality standards would enable products to be traded more competitively in international markets. It has been argued that less-developed countries could increase their share of the world fresh-produce market "if the necessary standards were met". Among these, food safety will be increasingly important in the future.

5.7 The role of agricultural engineers in postharvest development

As population density increases and land becomes scarce, decision there is increasing need to develop systems that allow this level of operation to work in a totally sustainable fashion and market the product, the agricultural engineer is trained to use his or her expertise to aid the industry in its development. Quality, safety, and reliability of product supply become primary if foreign trade and investment are to be secured and developed.

These are major issues that cannot be ignored. Agricultural engineers and scientists need to be aware of the challenges and then seek to find their own solutions that are best suited to the local culture and society. This may be far from the traditional concept of the designer and builder of machinery. He or she is also likely to be involved in the following.

Quality standards and management

Establishing quality standards is a key component of agricultural engineering practice significant long-term improvement in the quality of products. Export standards can be the main driving force, although it is possible for local industries to establish their own grading standards. Examples of standards include apple quality manual, which govern all aspects of fruit production including preharvest spraying, harvesting, time to cool after harvest, weight, and packaging requirements, as well as standards relating to colour, shape, and appearance.

Information transfer

Engineering initiatives in products packaging systems, such as modified-atmosphere storage, introduced new opportunities for extending the storage lifetime of products, but these require careful design and an understanding of the physiological aspects of storage. Even though postharvest researchers are becoming increasingly aware of the requirements for successful storage of commercially important food products, all too often systems are poorly designed or fail as a result of economic constraints, lack of knowledge, or inadequate quality management systems to ensure that standards are adhered to.

5.8 Future prospects and opportunities in postharvest development

Potentials of new crops

There are large numbers of crops, especially in less-developed countries, that bear edible fleshy parts or nuts of acceptable quality. These are largely harvested either from wild trees or seedlings nurtured in small plantings. The economic potential of these crops has been recognized, especially as traditional fruit crops come under severe pressure due to increasing production and declining profits.

Currently many such crops are produced only for domestic markets, where quality standards are often minimal or nonexistent. In any country, poor quality-control standards for produce result in unfavorable competition with foreign produce, even on the domestic market. For example, currently most less-developed countries suffer trade deficits on fruits and vegetables

Potentials of new technologies

The understanding and effective management of the process of technological change requires an approach involving the expertise and analytical skills of several disciplines from the engineering to the social sciences. It is also important to emphasize that less-developed countries should not be treated as a homogenous group, because particular strengths and weaknesses vary widely between countries.

CHAPTER 6

PRODUCTS DETERIORATION AND STORAGE SYSTEMS

Content: Horticultural crops deterioration, deterioration factors, post harvest storage conditions, fruits and vegetable packaging, packaging materials, fruit and vegetable storage, storages structures

6. Introduction

Deterioration commences at harvest; post harvest technologies are designed to slow the rate of ripening and senescence and hence quality decline. If deterioration is rapid, poor-quality product can be removed at the point of production or packing at which quality inspection occurs; if deterioration is slow the product may pass initial quality inspection yet be of reduced acceptability to consumers because of poor appearance, texture, and taste.

6.1 Factors affecting of products deterioration

Deterioration results from three main types of effects: physical, physiological, and pathological.

1. Physical factors

Produce can sustain mechanical or physical damage at all stages of the chain from harvest to consumption. Bruises, cuts, abrasions, and fractures occur as a result of poor handling or inadequate packaging. Such damage dramatically increases water loss and susceptibility to infection by post harvest fungi and bacteria. Product firmness and water status influence susceptibility to mechanical damage. Development of gentle yet effective handling systems and appropriate packages, together with education of personnel, are required to minimize physical damage.

2. *Physiological factors*

All living things respire to generate energy for continued metabolism. A simplified summary equation for respiration is:

$$C_6H_{12}O_6 + 6O_2 \rightarrow 6CO_2 + 6H_2O + heat\ energy \uparrow \cdots \ldots \ldots \ldots \ldots \ldots .6.1$$

Respiration rate of horticultural products varies, but as a general rule, perishability is a function of respiration rate; the greater the respiration rate the more perishable the product and the shorter the time it can be stored and still maintain acceptable quality. Actively growing products, such as asparagus, or some tropical fruits, such as mango, have high rates of respiration and limited storage life. Relative respiration rates of selected commodities are shown in Table 6-1 below

Table 6-1: Relative respiration rates of selected commodities:

Respiration rate	Commodity
Very low	Dates, dried fruits, nuts
Low	Apples, citrus, grapes, onions, mature potatoes, sweet potatoes
Moderate	Apricots, bananas, cabbage, carrots, cherries, mangoes, nectarines, peaches, pears, peppers, plums, immature potatoes, tomatoes
High	Avocados, blackberries, cauliflower, rasp berries, strawberries
Very high	Artichokes, brussel sprouts, cut flowers, green onions, snap beans
Extremely high	Asparagus, broccoli, mushroom, peas, spinach, sweet corn

Source: aplinfo.apl.com/reefer/html/knowing_your_cargo_chilled.html

Respiration is highly temperature dependent; the lower the temperature (down to 0 °C) of harvested fruit and vegetables the lower the respiration rate.

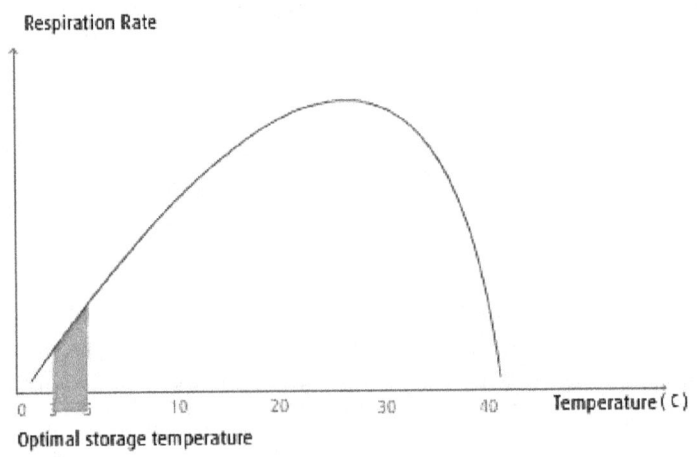

Figure 6-1: Temperature respiration relationship

Benefits of lowering respiration rate include:

a. Reduction in carbohydrate loss
b. Decreased rate of deterioration
c. Increased storage and shelf life

Low-temperature storage is the major weapon that the post harvest operators have to maintain quality and extend life of harvested products. Low temperatures not only reduce respiration rate, but also reduce water loss through transpiration and nutritional loss

Immediately after harvest, products should be placed in a well-ventilated shady environment to prevent large temperature increases that occur if products are exposed to direct sunlight. Rapid cooling as soon as possible after harvest (to reduce respiration rate) is used commonly in the horticultural industry, particularly for very perishable products such as strawberries and apples. Hydro cooling, vacuum cooling, or forced air-cooling is used widely to remove field heat rapidly. The method chosen depends on the crop.

3. *Pathological factors*

Any physical damage to product provides an ideal entry point for pathogens. A wide range of fungi and bacteria contribute to post harvest losses in fruit and vegetables throughout the world, especially in situations in which cool-chain management is inadequate.

These post harvest pathogens may infect produce at various preharvest stages through plant or fruit development, at harvest, or after harvest while products are in store or in transit to market. Removal of previously infected fruit and application of appropriate fungicides at bloom stage are the recommended means of control.

The severity of pathological infection can be measured in terms of a disease index based on a slightly modified scale of (0 - 5) developed by Frossard and Laville, 1973,

Where

0 = apparently healthy fruits
1 = slight infection
2 = 25 % of crown infected
3 = 50 % of crown infected
4 = 100 % of crown infected
5 = entire crown infected and infection progressing towards the pedicels

6.2   Factors affecting stored products deterioration

Under appropriate conditions, fruits can be stored for many years. It is important to establish the condition and storage period that afford the optimum balance between the cost of storage and the changes in quality of stored products. Almost all agricultural produce are stored before they are processed. Complex reactions do occur during such storage and the overall effect of these reactions manifest itself as quality change. The rate and the equilibrium of the reaction are influenced by many variables major of which are atmospheric conditions and ripeness.

Atmospheric conditions: The major components of atmospheric conditions are the temperature and humidity. Other important feature of the environment which influences the longevity of fresh produce is atmosphere surrounding the product. Wide range of atmospheric conditions exists in warehouses, dugouts, shelters and natural caves used for food storage. These can be highly satisfactory if conditions are sufficiently constant to allow prediction of storage life of the foods.

Figure 6-2: Fresh tomato fruits

Ripeness: Ripeness in fruits is often accompanied with softening which is seldom as serious as changes in colour or flavour in some products, if further processing is your goal. Slowing down ripening is beneficial for products meant for long storage. To slow down products ripening at home, you can wrap up fruits in polythene bags and store in the refrigerator. This will delay ripening

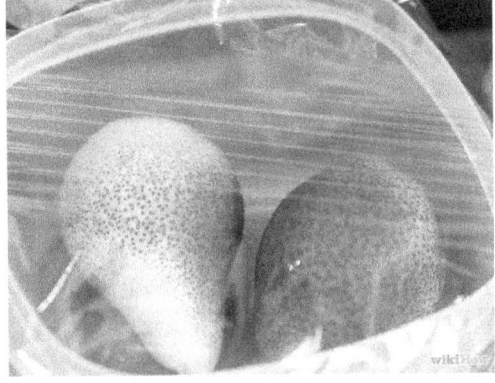

Figure 6-3: Storing pear in refrigerator

6.2.1. Influence of temperature

Temperature has by far the greatest effect on storage life but it is the easiest and most readily controlled. Deterioration in fruits is closely related to the natural ripening. The natural function of the fleshy parts of fruits is to deteriorate after the fruit leaves the tree and to release the seeds. Consideration that applies to the effects of temperature on fresh vegetables can often be applied to frozen products, although the temperature involved are typically $26°C$ lower. Loss of vitamins and nutritive values may be a more serious deterrent to long storage than loss of sensory appeal, particularly at temperatures in the $70°f$ to $32°f$ (21-0) $°C$ range.

The temperature maintained in the above ground natural storage houses vary with site and geographical location, from below freezing to above $100°f$ ($38°C$) at relative humidity range from almost 100% to 20% or below. Temperature in caves or deeper excavations may fluctuate at less than $10°F$ ($6°C$) annually. Utilization of these or similar sources of natural cool storage structures can yield advantages over above-ground storage in more uniform quality and increased life, or can reduce the cost of holding under refrigerated storage.

A higher rate of water loss will generally be experienced by:

- Leafy vegetables with larger exposed surface areas and in
- Plants that have been bruised or cut

Temperature management

Temperature management plays a key role in limiting water loss in storage and on transit. Transportation and display at roadside stands or farmer's markets often result in extended periods of exposure of sensitive produce to direct sun, warm (or even high) temperatures, and low relative humidity.

Figure 6-4: Roadside fruit markets

Rapid water loss under these conditions can result in limp, wilting, flaccid greens, softening, browning, stem separation and a loss of appealing natural sheen or gloss in fruits and vegetables or other defects.

Figure 6-5: Russeting, microcracks and moisture loss in apples

As the primary means of lowering respiration rates of fruits and vegetables, temperature has an important relationship to relative humidity and thus directly affects the product's rate of water loss. The relative humidity of the ambient air conditions in relation to the relative humidity of the crop (essentially 100%) directly influences the rate of water loss from produce at any point in the marketing chain.

By providing postharvest cooling before and during transport and a shading structure during display, you can minimize rapid water loss at these market outlets. During transportation and storage, relative humidity (more properly, *vapour pressure deficit*) is critical, even at a low temperature.

Water loss can be minimized by ensuring:

- Lowest safe temperature
- High humidity of 85%-95%
- Rapid pre-cooling
- Appropriate packaging

6.2.2. Influence of relative humidity

While temperature remain a primary concern in the storage of fruits and vegetables, relative humidity is also important. Humidity is not readily controlled, but can have a comparable effect on deterioration, not only does it regulate the rate of water loss, it causes physiological storage disorders. The relative humidity of the storage unit directly influences water loss in produce. Water loss can severely degrade quality — for instance, wilted greens may require excessive trimming, and grapes may shatter loose from clusters if their stems dry out. Water loss means salable weight loss and reduced profit.

The simplest method of increasing relative humidity of the storage air is to wet the floor of the room or mist the storage containers with cold water and allow the water to evaporate. For a more permanent system of high relative humidity in the storage environment, moisture can be added to the refrigerated air (Figure 6-6).

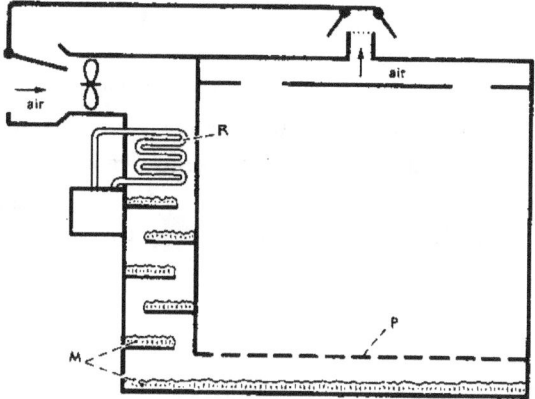

Figure 6-6: Wet moss as a moisture source inside a refrigerated storeroom

A fan draws air past the refrigerator's evaporator coils (R) then past wet moss or straw (M). The moist air is then pulled into the store-room through a perforated wall (P).

6.3. Postharvest storage systems

Fresh vegetables and fruits are living organisms and there is a continuation of life processes in them after harvest. To maintain the fresh vegetable in the living state, it is usually necessary to slow the life processes, avoiding death of the tissues, which produces gross deterioration and drastic difference in flavour, texture and appearance.

Changes in the harvested crops

Changes that occur in the harvested, unprocessed vegetable include

a. Water loss,
b. Conversion of starch to sugar, and vice versa,
c. Increase flavour changes,
d. Colour changes,
e. Toughening,
f. Vitamin gain or loss,
g. Sprouting, rooting,
h. Softening and decay.

While some changes results in quality deterioration, others improves quality in those vegetables that complete ripening after harvest. To maintain the fresh vegetable in the living state, it is usually necessary to slow down the life processes, though avoiding death of the tissues, which may produces gross deterioration and drastic difference in flavour, texture, and appearance. Vegetables for storage must be free from mechanical, insect, and disease injuries and should also be at the proper stage of maturity.

Methods of storage

Common (Unrefrigerated) storage and cold (refrigerated) storage are the methods generally employed for vegetable. Temporary storage, suitable for very brief storage periods, is frequently practiced in the shipping season when large lots are accumulated for car load or truck quantities. The refrigerator car or truck is a means of temporary storage while produce is in transit, short-term storage may last for four to six weeks.

Common (unrefrigerated) storage

Common storage lacks precise control of temperature and humility. Examples include the use of insulated storage houses, outdoor cellars or mounds. Cold storage allowed regulation of temperature and humidity and maintenance of constant conditions by use of a refrigeration and ventilation system.

Cold (refrigerated) storage

Refrigerated storage retards the following elements of deterioration in perishable crops:

- ⇨   Aging due to ripening, softening, and textural and colour changes;
- ⇨   Undesirable metabolic changes and respiratory heat production;
- ⇨   Moisture loss and the wilting that result;
- ⇨   Spoilage due to invasion by bacteria, fungi, and yeasts;
- ⇨   Undesirable growth, such as sprouting of potatoes.

One of the most important functions of refrigeration is to control the crop's respiration rate. Respiration generates heat as sugars, fats, and proteins in the cells of the crop are oxidized. The loss of these stored food reserves through respiration means decreased food value, loss of flavour, loss of salable weight, and more rapid deterioration. The respiration rate of a product strongly determines its transit and postharvest life. The higher the storage temperature, the higher the respiration rate will be.

For refrigeration to be effective in postponing deterioration, it is important that the temperature in cold storage rooms be kept as constant as possible. Exposure to alternating cold and warm temperatures may result in moisture accumulation on the surface of produce (sweating), which may hasten decay.

Storage rooms should be well insulated and adequately refrigerated, and should allow for air circulation to prevent temperature variation. Be sure that thermometers, thermostats, and manual temperature controls are of high quality, and check them periodically for accuracy.

Benefits of refrigerated storage systems

On-farm cooling facilities are a valuable asset for any produce operation. A grower who can cool and store produce has greater market flexibility because the need to market immediately after harvest is eliminated. The challenge, especially for small-scale producers, is the set-up cost. Innovative farmers and researchers have created a number of designs for low-cost structures.

Optimized refrigeration and controlled-atmosphere storage conditions reduce respiration, achieving slower deterioration and enhanced storage life. Preventing ethylene contamination of the storage environment from external sources or by ethylene producing crops is important to maximize storage life. These approaches, combined with new engineering and molecular approaches will be important in the production and handling systems used to get products to market.

6.4.   Postharvest storage conditions for selected fruits and vegetables

When buying fruit samples from loose fruit rack or open market, smell it and check for no bruising or mould or musty damp. Handle ripen fruit carefully, since ripe fruit spoils easily. Don't refrigerate a fruit before it ripens, since this ruins ripening. Once ripe, or immediately for non-ripening fruit, it is advisable to eat the fruit or otherwise,

1. Store in refrigerator to slow down respiration,
2. Store in plastic bag to stop moisture loss, or
3. Unsealed to avoid fermenting.

The following table presents list of conditions favourable to the picking and storage conditions of some certain fruits.

Fruit: Banana

Buying/ picking conditions: Buy green, so they won't easily be damaged .during transport.

Figure 6-7: Bananas

Storage condition: Ripen in paper bag until yellow with black specks. Then refrigerate what you can't eat.

Fruit: Pears/Avocado

Buying/ picking conditions: Buy pear under-ripe. Pick firm flesh with no blemishes. Avocado won't ripen on tree, because a chemical signal inhibits the process. When ripe, it yields to gentle pressure but skin is not loose.

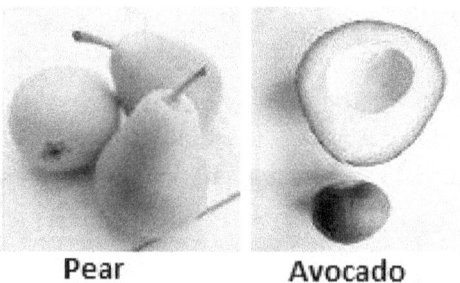

Pear   Avocado

Figure 6-8: Fleshy fruits

Storage condition: store pear in fridge (unusual practice) before ripening to avoid mushiness. Once ripe, their perfect ripeness lasts less than a day. Store avocado pears on tree! Or, after ripening, refrigerate it for up to a week.

Fruit: Drupe

Buying/ picking conditions: Best if picked after maturity. You can judge maturity by aroma. Apples smooth shiny skin, firm with no bruising, no blemishes. Papaya rich colour, gives slightly at room temperature.

Apple   Papaya   Mango

Figure 6-9: Drupes

Storage condition: (Apples lack polygalacturonase and so do not soften, instead remain crisp.)

Fruit: Plum and peaches

Buying/ picking conditions: Buy plum and peaches mature, or when they have fully-developed shoulders; sutures well-developed and just started to soften or when colour has no trace of green, or the nectarines have a slight sheen and no bruising.

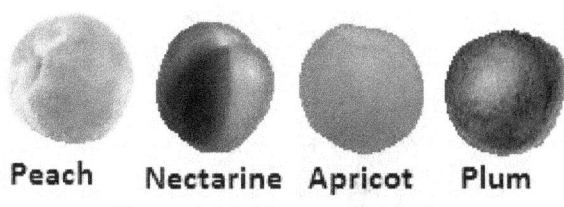

Peach   Nectarine   Apricot   Plum

Figure 6-10: Plum and peaches

Storage condition: Ripen in a sudden rush.

Fruit: Berries

Buying/ picking conditions: Buy mature fruits. Blueberries plump with blue/green "bloom", firm not soft.

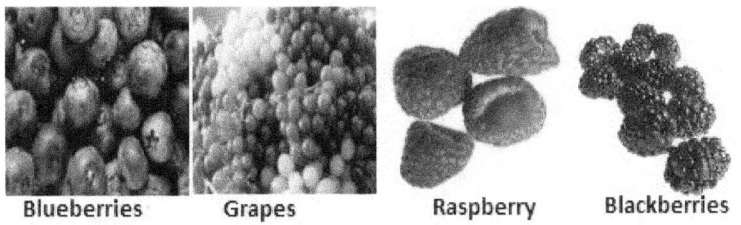

Blueberries   Grapes   Raspberry   Blackberries

Figure 6-11: Berries

Storage condition: Most ripen in a sudden rush. Berries have short storage life, therefore, store and handle berries carefully and briefly. Don't wash until last minute (to avoid damage/decay). Grapes keep okay in the fridge. Passion-fruit can be frozen.

Fruit: Melons

Buying/ picking conditions: Round depression where stem was should be smooth to show the fruit was ripe enough to fall off easily. Look for softness at opposite end. Cantaloupe skin is not green. Fruit with netted skin should have netting raised. With smooth skin, it should be slightly waxy.

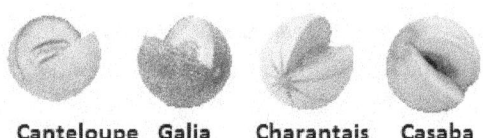

Canteloupe   Galia   Charantais   Casaba

Figure 6-112: Melons

Storage condition: Ripens in a sudden rush therefore, store and handle carefully and briefly.

Fruit: Watermelon

Buying/ picking conditions: Well-rounded at ends, rinds firm but not hard, intense red flesh without white streaks, seeds dark not white.

Figure 6-13: Watermelon

Storage condition: Ripens in a sudden rush therefore, store and handle carefully and briefly in cool storage. Avoid chilling atmosphere

Fruit: Orange/lemon/grapefruit

Buying/ picking conditions: Buy fruit that feels heavy for its size (will be juicier with more dissolved solids in juice). Look for glossy smooth even-coloured fine-pored skin (smooth skin implies thin skin). In oranges, button should be green, and green skin is fine.

Grapefruit   Lime   Orange   Tangelo

Figure 6-14: Citrus fruits

Storage condition: Store in cool refrigerated or room atmosphere; avoid chilling atmosphere

Fruit: Multiple fruits

Buying/ picking conditions: Choose fruits that already smell and taste ripe -- they will not ripen any further. Pineapple should have a strong sweet smell at its base. Choose fruits that already smell and tastes ripe - they will not ripen any further. Pineapple should have a strong sweet smell at its base.

Breadfruit   Jackfruit   Pineapple   Osage-orange

Figure 6-15: Multiple fruits

Storage condition: Has short storage lives. Don't wash until last minute (to avoid damage/decay). Pineapple will not get any *sweeter* in storage, but it will become *less tart* as its acids are used up to soften the flesh.

6.5. Fruits and vegetable packaging

Because it is possible to preserve food by altering their immediate environment, packaging has become an important element in preservation. Satisfactory packaging requires consideration of protection, economy, convenience and appearance. Factors affecting the choice of packaging materials include:

1. Product properties
2. Storage condition
3. Properties of economically available material.

Packaging is affected by the tendency of food to gain or lose moisture, its free or fat content, its particle size, tendency to sift and its susceptibility to spoilage by light, oxygen and organisms.

6.5.1. Properties of packaging materials

Material strength: Internal and external failures of both rigid and flexible containers frequently reduce quality or spoilage life of products, resistance to moisture, corrosion, leakage and package fatigue are needed to withstand high temperature or humidity or the corrosive action of certain products high in salt, fat natural acid sulphur compounds.

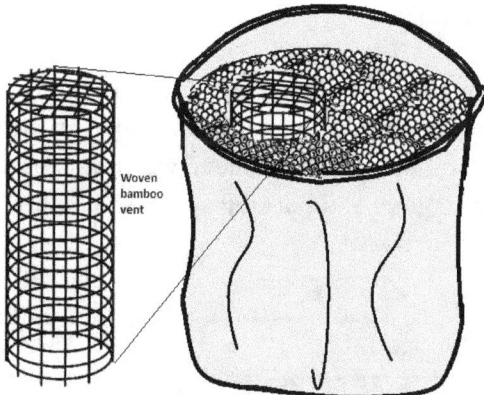

Figure 6-16: Vented bag of stored fruits

Air flow: Packages should be designed to allow cool air to flow directly over products, ensuring rapid temperature pull-down and consequent temperature maintenance. Regrettably, too many packages in current use are largely impenetrable to air movement, resulting in a slow temperature decrease, influencing quality negatively. If large bags or baskets must be used for bulk packaging of fruits or vegetables, the use of a simple vent can help reduce the build-up of heat as the product respires. In the illustration below, a tube of woven bamboo (about one meter long) is used to vent a large bag of fruit. Place the vent tube into the container before filling the container with product.

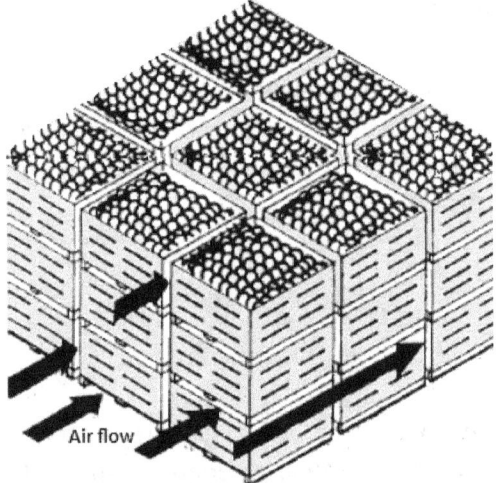

Figure 6-17: Pallet stacking pattern

Temperature control: Minimizing temperature variation from the optimum recommended during storage improves uniformity in product quality. Refinements are being made in refrigeration control systems, cool-store design, and pallet-stacking patterns to optimize air temperature and airflow through and around pallets in cool stores. This prevents hot spots from developing in localized areas within the store that would lead to higher respiration rates, more rapid deterioration, and hence poorer quality than in parts of the store in which product temperature was optimized.

Water/moisture loss: Packaging should also be designed to minimize water loss. To minimize condensation inside the bag and reduce the risk of microbial growth, the bags may be vented; micro perforated, or made of material permeable to water vapour. Barriers to water loss may also function as barriers to cooling, and packing systems should be carefully selected for the specific application with this in mind.

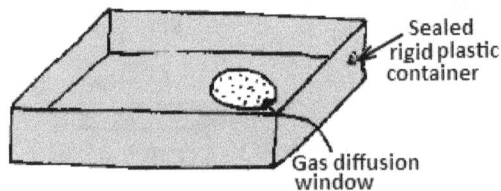

Figure 6-18: Vented plastic container

Chemical contamination: Packaging materials, storage or transport containers, or bins that contain synthetic fungicides, preservatives, or fumigants (or any bag or container that has previously been in contact with any prohibited substance) are not allowed for organic postharvest handling. In small-scale handling, the reuse of corrugated containers from conventional produce is strongly discouraged by organic certifying organizations.

Packaging conditions

Packaging containers exhibits two conditions of great significance in packaging. These are hermetic and non-hermetic conditions.

Hermetic condition: The term hermetic means a container which is absolutely impermeable to gases and vapours throughout its entirety as long as it remains closed, including its seams. This condition is impervious to bacteria, yeasts, moulds, and dirt from dust and other sources. The most common hermetic containers in use are rigid metal cans and glass bottles, glass containers etc.

Figure 6-19: Glass container for honey storage

Non-hermetic condition: A non-hermetic container is the one that prevents the entry of micro-organisms at all instance.

6.5.2. Fruits and vegetable packaging materials

Packaging containers

Common packaging containers are made from fibreboards or hard Kraft papers. Cartons can be glued, tape or stapled as desired during construction. Example of containers include one piece box, two-piece box with cover, full telescoping box, one-piece tuck-in box and interlocking box among several designs. Dimensions can be altered to suit the needs of the handler, and all containers should have adequate vents.

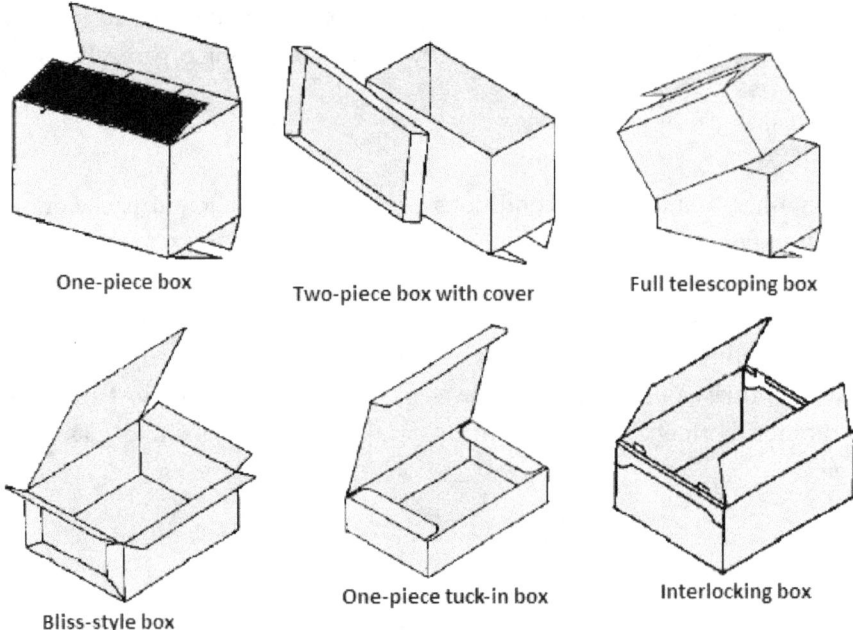

One-piece box   Two-piece box with cover   Full telescoping box

Bliss-style box   One-piece tuck-in box   Interlocking box

Figure 6-20: Fiberboard container designs

Primary and secondary containers

Primary containers by definition are those packaging materials (such as plastic liners) which come in direct contact with the food. Some foods are provided with efficient primary containers by nature, such as nuts, oranges, eggs etc.

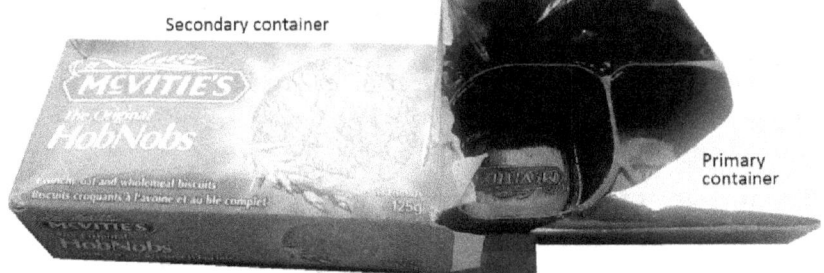

Figure 6-21: Primary & secondary container

Secondary containers: Secondary containers are those protective cartons or drums for packaging materials filled into primary containers such as plastic liners. Foods such as milk, dried eggs and fruit concentrates are often packaged within secondary containers. In packaging these, we generally need only a secondary outer box, wrap, or drum to hold units together and give gross protection.

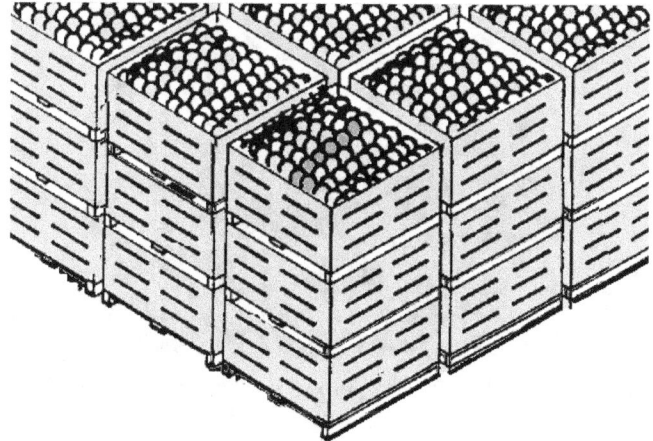

Figure 6-22: Secondary containers arranged in stacks

Waxed cartons, wooden crates and plastic containers:

If produce is packed for ease of handling, waxed cartons, wooden crates or rigid plastic containers are preferable to bags or open baskets, since bags and baskets provide no protection to the produce when stacked. These materials are reusable and cost effective when used for the domestic market and can also withstand high relative humidity found in the storage environment.

Figure 6-23: Carton with fiberboard divider

Plastic field boxes

These types of boxes are usually made of polyvinyl chloride, polypropylene, or polyethylene materials. They are durable and can last many years. Many are designed in such a way that they can nest inside each other when empty to facilitate transport, and can stack one on top of the other without crushing the fruit when full (Figure 6-24).

Figure 6-24: Plastic field boxes with nest/stack design

Wooden field boxes

These boxes are made of thin pieces of wood bound together with wire. They come in two sizes: the bushel box with a volume of 2200 in³ (36052 cm³) and the half-bushel box. They are advantageous because they can be packed flat and are inexpensive, and thus could be non-returnable. They have the disadvantage of providing little protection from mechanical damage to the produce during transport. Rigid wooden boxes of different capacities are commonly used to transport produce to the packinghouse or to market. (Figure 6-25).

Figure 6-25: Typical wooden crate holding fresh tomatoes.

Bulk bins

Bulk bins are made of wood and plastic materials of capacity between 200-500 kg and are used for harvesting fresh fruits and vegetables. These bins are much more economical than the field boxes, both in terms of fruit carried per unit volume and durability, as well as in providing better protection to the product during transport to the packinghouse.

Figure 6-26: Field bins

Hand-pack

Produce can be hand-packed to create an attractive pack, often using a fixed count of uniformly sized units. Plastic crates (Figure 6-27) can be used many times, reducing the cost of transport. They are available in different sizes and colours and are resistant to adverse weather conditions. However, plastic crates can damage some soft produce due to their hard surfaces, thus liners are recommended when using such crates.

Figure 6-27: Plastic crate of with oranges

Packaging materials such as trays, cups, wraps, liners and pads may be added to help immobilize the produce. Simple mechanical packing systems often use the volume-fill method or tight-fill method, in which sorted produce is delivered into boxes, vibrated and then settled are shown in Figure 6-28 below.

Figure 6-28: Samples packaging of apples with RPC and DRC plastics

Films and foils

Films and foils have different values for moisture and gas permeability, strength, elasticity, inflammability and resistance to insect penetration and many of these characteristics depend upon the film's thickness. Films are used in the construction of inner containers. Since they are non-rigid, their main functions are to contain the product and protect it from contact with air or water vapour. Their capacity to protect against mechanical damage is limited, particularly when thin films are considered. Common examples of films are paper, aluminum foil polythene, cellulose acetate etc

Properties of these materials include:

1. Negligible permeability to water-vapour, gases and odours;
2. Grease proof and printable,
3. Opaque and brilliant/glossy appearance;
4. Good dimensional stability;
5. Dead folding characteristics.

Plastic sheets

Typical examples of plastic sheet packaging materials include;

- *Cellophane paper.* This can be used for packing dried products, mainly for dried fruit leathers.
- *Polyethylene sheets* have a variety of uses. They are flexible, transparent and have a perfect resistance to low temperatures and impermeability to water vapour. An important advantage is that these sheets can be easily heat-sealed

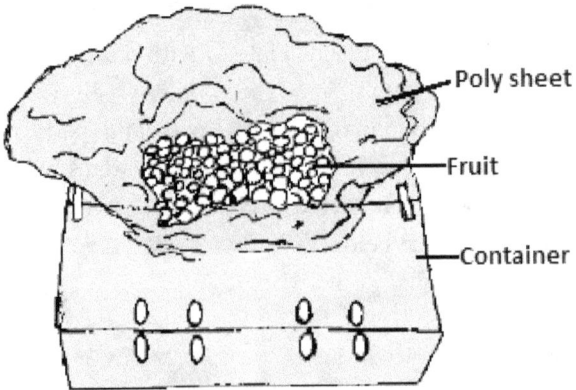

Figure 6-29: Glass receptacle flask

Various flexible materials such as papers, plastic films, and thin metal foils have different properties with respect to water vapour transmission, oxygen permeability, light transmission, burst strength, pin holes and crease hole sensitivity, etc.

Figure 6-30: Sealed plastic bag

Many plastic films are available for packaging, but very few have gas permeability that makes them suitable for MAP. Low density polyethylene and polyvinyl chloride are the main films used in packaging fresh fruits and vegetables. Polyesters have low gas permeabilities that they are suitable only for commodities with very low respiration rates.

Pressure-pack trays

Once sorted into grades and size, fruit are delivered by conveyor to a packing area. Here fruit may be held in rotating final size bins until they can be packed by hand into cartons. Alternatively, if automatic tray fillers are used, fruit fall into a paper pulp tray directly. Fruit and vegetables also may be wrapped individually in foam liners or films before, or instead of, packing them in a rigid container.

Figure 6-31: Pressure-pack tray system

Peg-type trays

Packaging fulfils several functions including containment, facilitating transportation, protection of fruit from further damage, protection of the environment from contents of package (for example, if the contents are dirty), marketing, product advertising, and stock control.

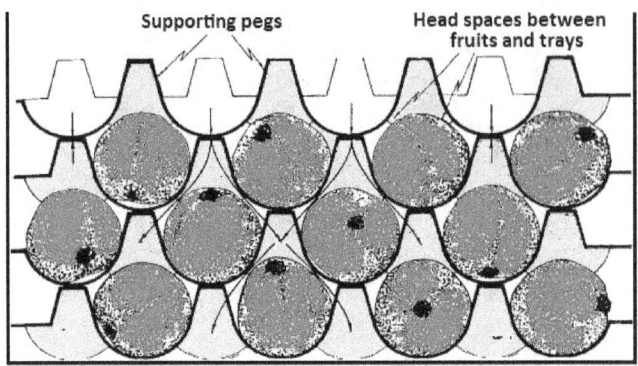

Figure 6-32: Peg-type tray system

Receptacles and packaging in plastic materials

There are three categories of plastic receptacles in common use in fruit packaging:

a. *Heat resistant receptacles:* These are cellophane or *polyethylene* receptacle box, bottle and bags that can be heat treated up to 120 °C. Polyethylene bags could be used to some extent for packing and pasteurization of sauerkraut.
b. *Non-heat resistant receptacles: These are* receptacles that are not heat treated during processing of fruit and vegetables, also available in bags and boxes. Bags are the most used type of packing from plastic materials and they are manufactured from polyethylene or cellophane; an important utilisation is for dried/dehydrated fruits and vegetables.
c. *Special packaging* - which can be contacted (Criovac type) by action of heat once the finished product is already inside the pack and the air is evacuated.

Paper packaging

As primary containers few paper products are not treated, coated or laminated to improve their protective properties. Paper from wood pulp and reprocessed waste paper will be bleached and coated or impregnated with such materials as waxes, resins, lacquers, plastics, and laminations of aluminium to improve water vapour and gas impermeability, flexibility, tear resistance, burst strength, wet strength, grease resistance, sealability, appearance, printability, etc.

Kinds of paper sheets

The following kinds of paper are commonly used in packaging

- *Kraft paper* is the brown unbleached heavy duty paper commonly used for bags and for wrapping; it is seldom used as a primary container;

Figure 6-33: Various paper sheet packaging

- *Parchment paper*: acid treatment of paper pulp modifies the cellulose and gives water and oil resistance and considerable wet strength to this type of packaging material;
- *Glassine-type papers* are characterized by long wood pulp fibers which impart increased physical strength;
- *Paper with plastic material sheets.*

Figure 6-34: Plastic/paper combination container for flowers

Glass containers

Glass is a chemically inert food packaging material, although the usual problems of corrosion and reactivity of metal closures will of course apply. The principal limitation of glass is its susceptibility to breakage, which may be from internal pressure, impact, or thermal shock, all of which can be greatly minimized by proper matching of the container to its intended use and intelligent handling practices.

Main classes of glass receptacles are:

a. *Jars which are resistant to heat treatments*: These receptacles may replace metal cans considering that they do not react to food content, transparent, reusable and can be manufactured in various shapes. Perfect hermeticity after heat treatment (pasteurization/sterilization) and cooling may be achieved by the use of metallic (or glass) caps and specific materials for tightness.

b. Jars for products without heat treatment: Jars, glasses, etc. For products not submitted to heat treatment (marmalades, acidified vegetables, etc.), The jars can be filled hot and therefore sterile in receptacles, pneumatically closed to protect it against negative air action which is in this case eliminated from receptacles. The use of jars with pneumatic closure presents the advantage that some products (e.g. marmalades, jams) can be filled hot and therefore sterile in receptacles.

Figure 6-35: Glass receptacle flask

c. *Glass bottles*: These receptacles are widely used both finished products which need pasteurization such as tomato juice, fruit juices, etc. and for those which are preserved non-pasteurized (syrups) and

d. *Receptacles with higher capacity* (flasks, etc.): These are glass flasks with 3 to 10 liter capacity which are resistant to product pasteurization and can be hermetically closed (e.g. tomato juice).

Tin can/tinplate

The "tin can" is a container made of tinplate. *Tinplate*, a rigid and impervious material, consists of a thin sheet of low carbon steel coated on both sides with a very thin layer of tin. It can be produced by dipping sheets of mild steel in molten tin (hot-dipped tinplate) or by the electro-plating of tin on the steel sheet (electrolytic tinplate). With the latter process it is possible to produce tinplate with a heavier coating of tin on one surface than the other (differentially coated).

Tin is not completely resistant to corrosion but its rate of reaction with many food materials is considerably slower than that of steel. The effectiveness of a tin coating depends on:

a. Its thickness which may vary from about 0.5 to 2.0 μm (20 to 80 x 10(-6) in.);

b. The uniformity of this thickness;

c. The method of applying the tin which today primarily involves electrolytic plating;

d. The composition of the underlying steel base plate;

e. The type of food, and

f. Other factors.

Tinplate sheets may be lacquered after fabrication to provide an internal or external coating to protect the metal surface from corrosion by the atmosphere or through reaction with the can contents.

6.6. Effect of storage and packaging on food

Colour stability varies widely among different products and storage temperature. The only significant change at (0-20°C) (-18-29°C) is a slightly lighter appearance, and there is no significant change in any product at 32 °F (0 °C). Temperature of 47°f (8°C) could result in significant changes; particularly in the stable items.

At temperatures from (70-100) °f (21-30) °C, products tend to darkened, brown, fade and lose flavor or become stale. Extremes of temperature have adverse effects on texture. Freezing causes sufficient damage to such items as tomatoes and beans, thus reduce them to substandard grades.

6.7. Storage of horticultural crops

If produce were to be stored, it is important to select high quality products with sound and wholesome crops. Products must be screened to avoid the presence of damaged or diseased units, and containers must be well ventilated and strong enough to withstand stacking. In general, proper storage practices include temperature control, relative humidity control, air circulation and maintenance of space between containers for adequate ventilation, and avoiding incompatible product mixes. Commodities stored together should be capable of tolerating the same temperature, relative humidity and level of ethylene in the storage environment.

 High ethylene producing crops (such as ripe bananas, apples, cantaloupe) can stimulate physiological changes in ethylene sensitive commodities (such as lettuce, cucumbers, carrots, potatoes, sweet potatoes) leading to often undesirable colour, flavour and texture changes.

Temperature management during storage can be aided by constructing square rather than rectangular buildings. Rectangular buildings have more wall area per square feet of storage space, so more heat is conducted across the walls, making them more expensive to cool. Temperature management can also be aided by shading buildings, painting storehouses white to help reflect the sun's rays, or by using sprinkler systems on the roof of a building for evaporative cooling. The United Nations' Food and Agriculture Organization (FAO) recommends the use of ferrocement for the construction of storage structures in tropical regions, with thick walls to provide insulation.

Low cost cold rooms can be constructed using concrete for floors and polyurethane foam as insulation materials. Building the storeroom in the shape of a cube will reduce the surface area per unit volume of storage space, also reducing construction and refrigeration costs.

6.7.1. Storage practices

The following practices are best suited for safe storage of horticultural crops.

1. Regular inspection of stored produce and cleaning storage structures on a regular basis will help reduce losses by minimizing the build-up of pests and discouraging the spread of diseases.
2. Storage facilities should be protected from rodents by keeping the immediate area clean, free from trash and weeds.

Figure 6-36: Regular inspection of stored produce and cleaning

3. Rat guards can be made from simple materials such as old tin cans or pieces of sheet metal fashioned to fit the extended legs of storage structures.

Figure 6-37: Rat guards

4. Reusable containers and sacks should be disinfected in chlorinated or boiling water before reuse.

Figure 6-38: Disinfecting reusable containers

5. Storing sacked products on platform sheets or raised platforms (wooden pallets) prevents dampness from reaching produce suited to dry conditions in storage. This helps to reduce the chance of fungal infection, while also improving ventilation and/or sanitation in the storeroom.

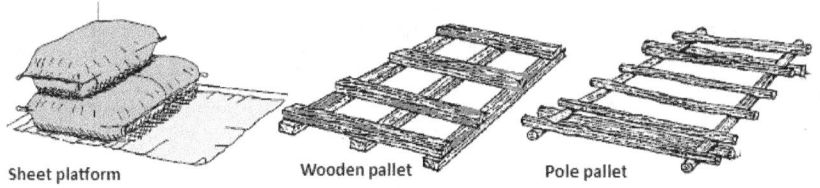

Sheet platform   Wooden pallet   Pole pallet

Figure 6-39: Storing sacked products on platforms

6. Concrete floors will help prevent rodent entry, as will screens on windows, vents and drains.

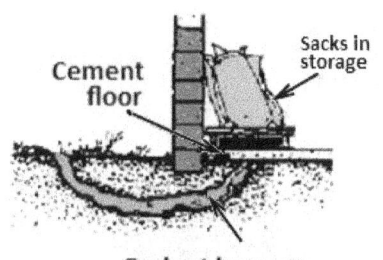

Figure 6-40: Concrete floor to prevent rodent entry

6.7.2. Fresh vegetable storage

Vegetables can be stored in fresh state under some specific natural conditions without significant modifications of their initial organoleptic properties. Fresh vegetable storage can be short term; this was briefly covered under temporary storage before processing. Also fresh vegetable storage can be long term during the cold season in some countries and in this case it is an important method for vegetable preservation in the natural state.

In order to assure preservation in long term storage, it is necessary to reduce respiration and transpiration intensity to a minimum possible; this can be achieved by:

a. Maintenance of as low a temperature as possible (down to 0° C),
b. Increase air relative humidity up to 85-95 % and
c. Increase CO_2 percentage in air related to the vegetable species.

From the time of harvest and throughout the period of storage, vegetables are subjected to respiration and transpiration and this is on account of their reserve substances and water content. The more the intensity of these two natural processes are reduced, the longer the storage time will be with reduced losses.

Some optimal storage conditions for some tropical fresh vegetable are shown in table 6-1 below

Table 6-1: Optimal conditions for fresh vegetable storage

Vegetables	Storage conditions	
	Temperature, °C	Relative humidity, %
Potatoes	+1- +3	85-90
Carrots	0 - +1	90-95
Onions	0 - +1	75-85
Leeks	0 - +0.5	85-90
Cabbage	-1 - 0	90-97
Garlic	0 - +1	85-90
Beets	0 - +1	90-95

Source: FAO 1995

6.8. Storage structures for horticultural crops

Requirements for storage structures for fruits and vegetable

Any building or facility designed for storage of horticultural crops should be conformable with the requirements for safe storage standards. Some of the requirements for safe storage structures include:

Insulation: Crops storage structures must be insulated for maximum effectiveness. A well insulated refrigerated building will require less electricity to keep produce cool. If the structure is to be cooled by evaporative or night air ventilation, a well insulated building will hold the cooled air longer. Insulation R-values are listed below for some common building materials. R refers to resistance and the higher the R-value, the higher the material's resistance to heat conduction and the better the insulating property of the material.

Table 6-2: R-value for insulation materials

Material	1 inch thick
Batt and Blanket Insulation	
Glass wool, mineral wool, or fibreglass	3.50
Fill-type insulation	
Cellulose	3.50
Glass or mineral wool	2.50-3.00
Vermiculite	2.20
Wood shavings or sawdust	2.22
Rigid insulation	
Plain expanded extruded polystyrene	5.00
Expanded rubber	4.55
Expanded polystyrene moulded beads	3.57
Aged expanded polyurethane	6.25
Glass fiber	4.00
Polyisocyranuate	8.00
Wood or cane fiber board	2.50
Foamed-in-place Insulation	
Sprayed expanded urethane	6.25

Source: Kitinoja and Adel, 2002

Low cost cold rooms can be constructed using concrete for floors and polyurethane foam as an insulator. Illustrated below is a cross-sectional view of a storehouse for fruits. This system was officially approved as the standard model for farm-level storehouses by the Ministry of Construction (Korea) in 1983. Note that air inlets are at the base of the building, and the floor is perforated, allowing free movement of air. The entire building is set below ground level taking advantage of the cooling properties of soil.

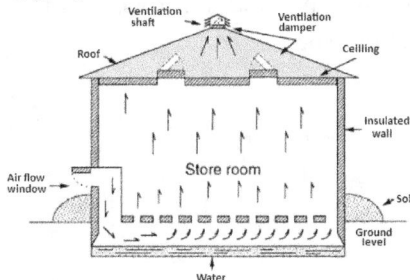

Figure 6-41: Cross-sectional view of a storehouse for fruits (Seung Koo Lee, 1994)

Storage using refrigerated facilities requires no further ventilation. For these systems, a simple recirculation system can be designed by adding a fan below floor level and providing a free space at one end of the storeroom for cool air to return to the inlet vents.

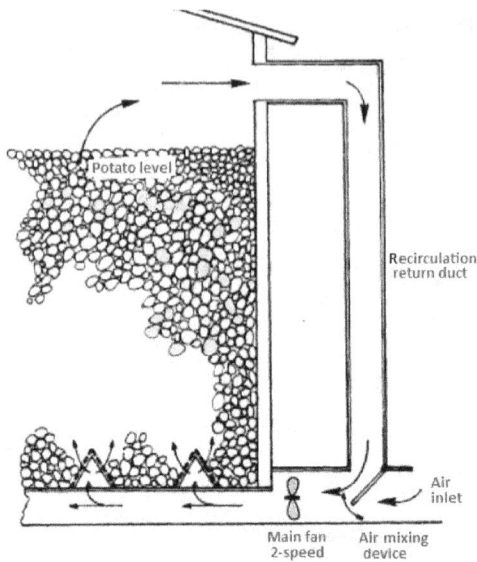

Figure 6-42: Cross-sectional view of a refrigerated storehouse (Potato Marketing Board)

Lateral ducts can be constructed of a variety of materials. Portable vents can be made from wooden slats, in a triangular, square or rectangular design. A round tube of plastic or clay can be used if holes can be drilled without causing structural damage, or permanent ducts can be constructed below ground, using concrete blocks.

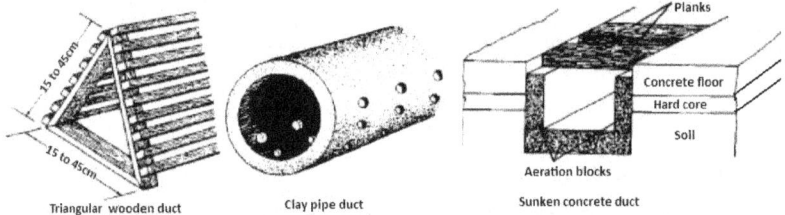

Figure 6-43: Lateral ducts

Overhanging roof extensions on storage structures are very helpful in shading the walls and ventilation openings from the sun's rays, and in providing protection from rain. An overhang of at least 1 meter (3 feet) is recommended.

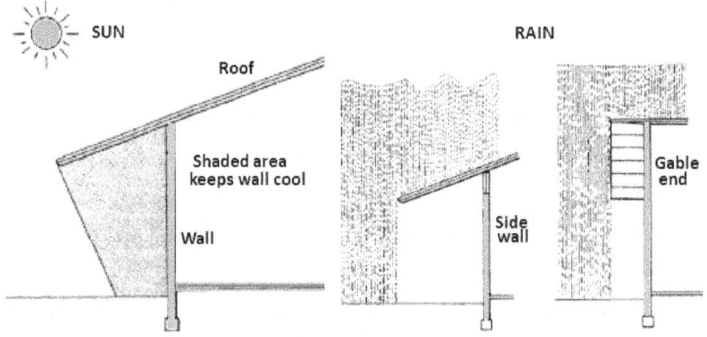

Figure 6-44: Overhanging roofs and shades (Walker, 1992)

Surface storage

Protected surface storage is a simple method for storing small quantities of produce. Insulating materials such as straw can be used and protective covers can be constructed from wooden planks, plastic sheeting or layers of compacted soil. Onions require a cool and dry atmosphere; and pumpkins, squash and sweet potatoes need a dry place where it is relatively warm, as such requires surface storage. Onions, after being pulled, should be cured by spreading them in the sun for several days. Remove tops three or four inches above the bulb after curing. Place on slatted racks or trays and store in a cool, dry place where there will be circulation of air.

Figure 6-45: Surface storage

Pumpkins and squash should be handled carefully to prevent bruising and stored on shelves in a dry room where the temperature will be about 50 degrees F. If warmer they will lose weight; if moist they will rot. Sweet potatoes need the same conditions as pumpkins and squash.

Pit and mound storages

Cold-requiring vegetables can be stored in basement cold room, or an outdoor cellar; where not available, the alternative is to store them in pits or mounds outdoors. Cabbage, turnips, radishes, beets, carrots, Brussels sprouts, celery, potatoes and apples may be stored in cool and moist surroundings.

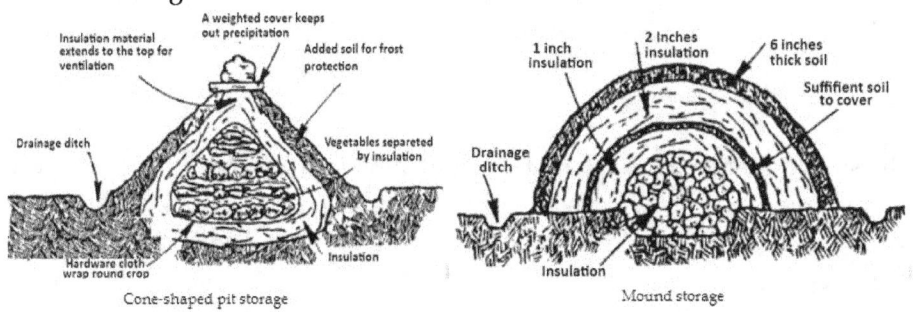

Figure 6-46: Cone-shaped pit and mound storage (Walker, 1992)

Trench storage

Vegetables such as beets, carrots and turnips can be stored in boxes between layers of loose soil or sand in well drained dug-out trenches. Make the trench deep enough to leave the tops of the boxes a foot below ground level, and large enough to hold all the boxes; place the boxes in the trench, lay boards across the top of the trench, stand a piece of drain pipe or tile up to carry off the air and after a few days when the vegetables have cooled, throw in enough earth to close the space between the roots and the boards.

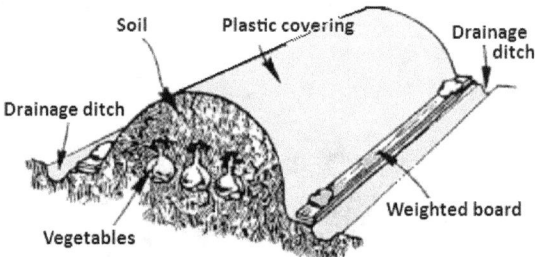

Figure 6-47: Trench storage (Walker, 1992)

Potatoes may be stored in this way if plenty of straw is used so air circulates around them (Figure 6-48). For cabbage, a simple method is to dig a trench about 8 inches deep and wide enough for three heads (Figure 6-48).

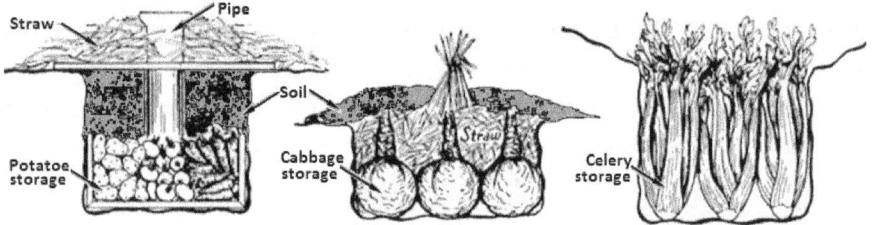

Figure 6-48: Trench storage of vegetables

Pull the heads up by the roots, remove the largest outer leaves, and place the heads top down in the trench. Cover with straw or hay, then soil, and add more soil as the weather becomes colder. Stand a bunch of straw up like a chimney every few feet along the trench to give ventilation.

Storage barrel

One of the simplest methods for storing small quantities of produce is to use any available container, and create a cool environment for storage by burying the container using insulating materials and soil. The example provided here employs a wooden barrel and straw for insulation.

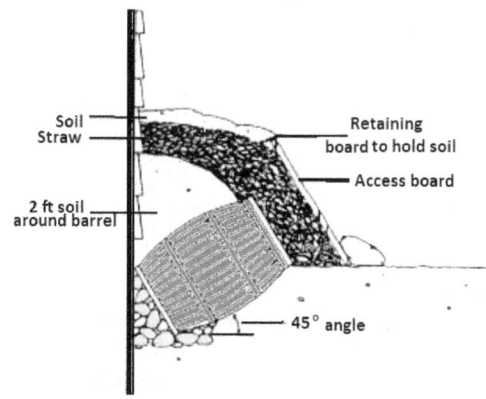

Figure 6-49: Storage barrel (McKay, 1992)

Storage bin

A root box, lined with hardware cloth and straw, buried to the top edge in soil will keep potatoes cool while providing protection from freezing. The wooden lid can be lifted for easy access to produce, and straw bales on top provide more insulation.

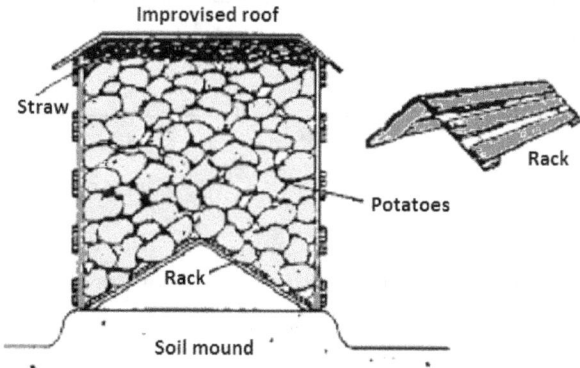

Figure 6-50: Storage bin (Bubel and Bubel, 1979)

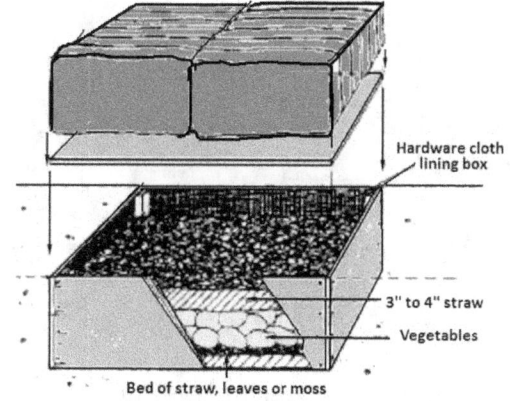

Figure 6-51: Storage bin (Bubel and Bubel, 1979)

Storage of dried and bulb crops

Onions, garlic and dried produce are best suited to low humidity in storage. Onions and garlic will sprout if stored at intermediate temperatures. Pungent types of onions have high soluble solids and will store longer than mild or "sweet" onions, which are rarely stored for more than one month. The following table lists the storage conditions recommended for these crops.

Table 6-3: Storage conditions recommended for dried and bulb crops

	Temperature		Relative Humidity	Potential storage duration
	°C	°F	%	
Onions	0-5	32-41	65-70	6-8 months
	28-30	82-86	70	1 month
Garlic	0	32	70	6-7 months
	28-30	82-86	65-70	1 month
Dried fruits and vegetables	<10	<50	55-60	6-12 months

Bulk storage

For fruits that can be stored in bulk such as onions or garlic, ventilation systems should be designed to provide air into the store from the bottom of the room at a rate of 2 cubic feet per minute per cubic feet of product.

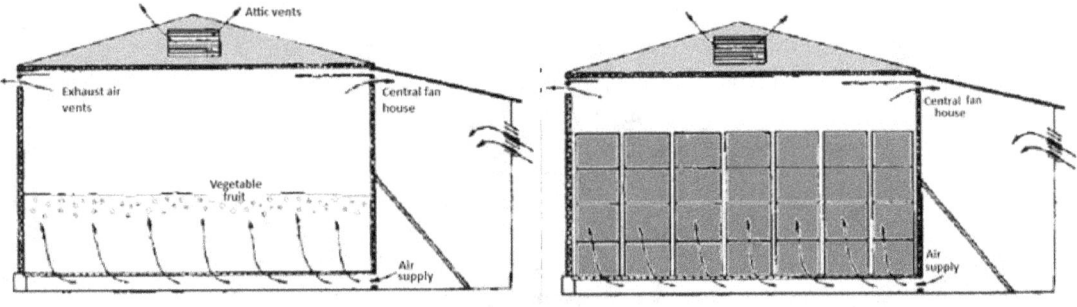

Figure 6-52: Bulk storage (Cantwell and Kasmire. 2002)

Storage in cartons or bins

If produce is in cartons or bins, stacks must allow free movement of air. Rows of containers should be stacked parallel to the direction of the flow of air and be spaced six to seven inches apart. An adequate air supply must be provided at the bottom of each row and containers must be properly vented.

Modified atmosphere (MA) storage

In view of increased globalization of food market and trans-border trades, modified atmosphere storage technology; a practice in which the internal atmospheric condition is passively regulated depending on product mass, temperature, and nature of a polymeric film package, being used has become commonplace. Typical examples include food packages, drugs, fruits etc. The modification process often tries to lower the amount of oxygen (O_2), moving it from 20.9% to 0%, in order to slow down the growth of aerobic organisms and the speed of oxidation reactions (Church and Parson, 1995).

Figure 6-53: Pears packed in modified atmosphere primary storage bags

Modified atmosphere storage technology has been proven most suitable for preserving natural quality of food products and extension of the products' storage life.

Controlled atmosphere (CA) storage

Controlled atmosphere storage or "CA" storage refers to a non-chemical process of storage of products under regulated atmosphere. In CA storage, the oxygen level is reduced and the carbon dioxide level is increased. Generally controlled-atmosphere stores operate with ±0°C atmosphere containing 1% to 5% CO_2 and 1% to 3% O_2, depending on crop and cultivar. In addition, they can be used to reduce disorders in whole fruit.

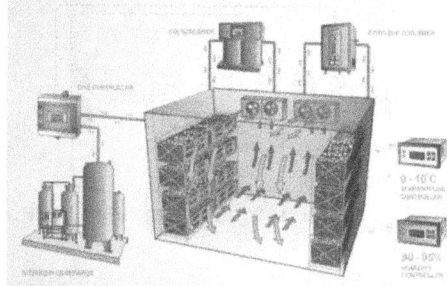

Figure 6-54: Controlled atmosphere mechanism (Agroripe.com)

Controlled or modified atmosphere storage should be used as a supplement to, and not as a substitute for, proper temperature and relative humidity management. Air coming into the storeroom or being re-circulated within the room must pass through a monitoring and control system.

6.9. Transportation of horticultural crops

Horticultural crops especially fruits and vegetables are moved from one location to another for the purposes of even distribution of crops, overcoming seasonal variation, and regional nativity and for of further processing, market delivery among others. It is however important that produce leaving packing facility must be suited to the handling it will receive as it is transported to market. Critical elements of horticultural crops transport include; the maturity stage of the crop, ripeness and extent of mechanical damage.

Requirements for crop transport

The requirements for products meant for transportation include:

Temperature management is critical during long distance transport, so loads must be stacked to enable proper air circulation to carry away heat from the produce itself as well as incoming heat from the atmosphere and off the road.

Transport vehicles should be well insulated to maintain cool environments for pre-cooled commodities and well ventilated to allow air movement through the produce. During transport, produce must be stacked in ways that minimize damage, and then be braced and secured.

Adequate air circulation: Produce transported in cartons should also be stacked so as to allow adequate air circulation throughout the load.

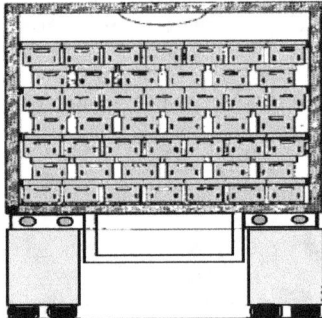

Figure 6-55: Well stacked container (Cantwell and Kasmire. 2002)

When cartons of various sizes must be loaded together, the larger, heavier containers should be placed on the bottom of the load. Parallel channels should be left for air to move through the length of the load.

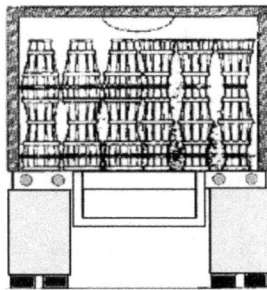

Figure 6-56: Well stacked crates (Cantwell and Kasmire. 2002)

Containers should be loaded so that they are away from the side walls and the floor of the transport vehicle in order to minimize the conduction of heat from the outside environment. There should always be a void between the last stack of produce and the back of the transport vehicle.

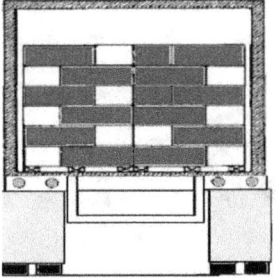

Figure 6-57: Well stacked containers (Cantwell and Kasmire. 2002)

The load should be braced to prevent shifting against the rear door during transit. If the load shifts, it can block air circulation, and fallen cartons can present great danger to workers who open the door at a destination market. A simple wooden brace can be constructed and installed to prevent damage during transport

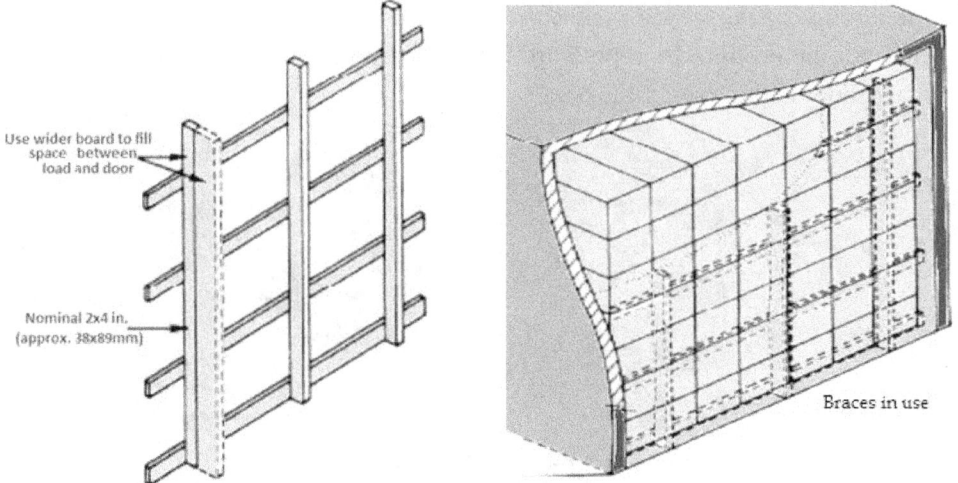

Figure 6-58: Wooden braces in use (Cantwell and Kasmire. 2002)

Types of transport systems

The following types of transport systems are common with fruits and vegetable transport

1. Open vehicles

Bulk loads of produce should be carefully loaded so as not to cause mechanical damage. Vehicles can be padded or lined with a thick layer of straw. Woven mats or sacks can be used in the beds of small vehicles. Other loads should not be placed on top of the bulk commodity. A truck ventilating device can be constructed for an un-refrigerated open vehicle by covering the load loosely with canvas and fashioning a wind catcher from sheet metal.

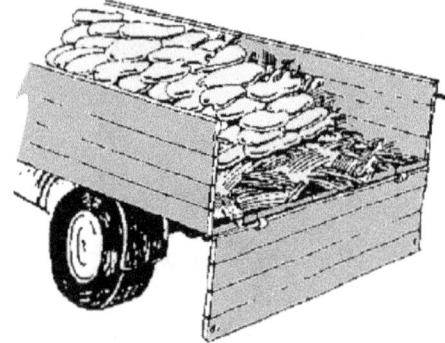

Figure 6-59: Wooden brace (Cantwell and Kasmire. 2002)

2. Portacoolers

Portacoolers are mobile precooling vans designed to extract heat as quickly as possible and desirable by using forced air flow through the load to be transported. A typical precooler is the USDA portacooler. This small cooler uses a 12,000 BTU/hr (1 ton) 110 volt room window air conditioner to cool air inside the insulated box. The cool air inside the front of the box is forced through the produce by a pressure fan in a second wall inside. The return air passes under a false floor to the front of the box.

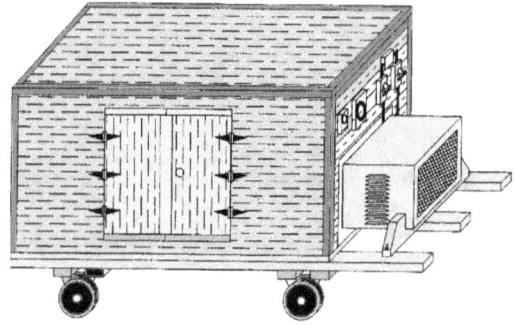

Figure 6-60: USDA Portacooler

The USDA post harvest cooling program stresses to small growers the need to:

a. Sort and grade produce out of the field,
b. Package the produce properly for the market,
c. Immediately cool the produce to remove field heat.

3. *Refrigerated trailers*

For optimum transport temperature management, refrigerated trailers need insulation, a high capacity refrigeration unit and fan, and an air delivery duct. The condition of the inside of a refrigerated trailer affects its ability to maintain desired temperatures during transport. Handlers should inspect the trailer before loading, and check these features:

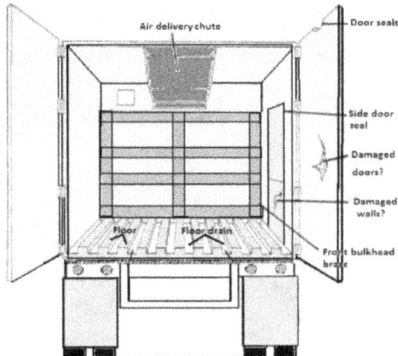

Figure 6-61: Refrigerated trailers

The condition of the inside of a refrigerated trailer affects its ability to maintain desired temperatures during transport. Handlers should inspect the trailer before loading, and check these features:

a. Clean floor,
b. Open drain floor
c. Installed front bulkhead brace
d. Inspects walls and doors for any damage and possible repairs
e. Inspect side door and door seals for tightness
f. Check air delivery chute to ensure its being intact

4. *Air transport*

Air transport became popular in view of trans-continental trade agreements and the need for easy distribution of climacteric crops to across the world. Air transport of fruit products requires a lot of handling and organization.

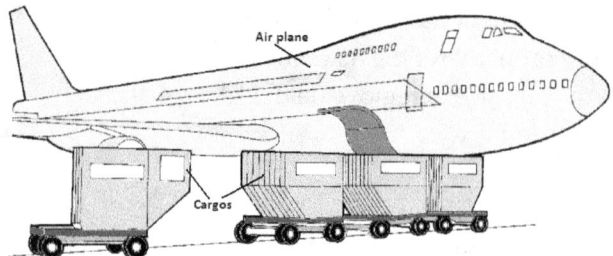

Figure 6-62: **Air** transport (McGregor, 1987)

To prevent shifting of the load in a cargo container for air transport, pieces of solid foam or folded fiber-board or associated packaging materials should be placed along the curved or triangular portion of the floor of the container. Cartons stacked on top will be much better supported and be held upright.

CHAPTER 7

TECHNOLOGICAL PROCESSES FOR FRUITS AND VEGETABLES

Content: Quality of fruits, quality measurement and determination, technological processes for fruits and vegetable, fruit and vegetable preservation technologies, challenges and future prospects of fruit technology processing

7.   Introduction

In developing countries, agriculture is the main stay of their economy. Of the various types of activities that can be termed agricultural based, fruit processing is one of the most important. In the processing of fruits, the quality of the fruit, the picking method and quality measurement are very important.

7.1   Objectives of fruit and vegetable technological processes

Many technologies are being adopted for the processing of fruits and its associated products. The main objective of these processing technologies is to provide a suitable condition in the fruit for preservation purposes.

Fruit processing is aimed at supplying wholesome, safe, nutritious and acceptable food to consumers throughout the year. Established fruit processing projects are aimed at solving clearly identified developmental problems due to:

a.   Insufficient demand
b.   Weak infrastructures.
c.   Poor transportation facilities and
d.   Perishable nature of the crops.

Processing plants are established in developing countries for the following reasons:

1. Diversification of the economy in order to reduce present dependence on one export commodity.
2. Develop new value added products.
3. Reduce fruit and vegetable losses.
4. Government industrialization policy.
5. Reduction of imports and meeting export demands.
6. Stimulating agricultural production by obtaining marketable products.
7. Generate both rural and urban development
8. Improve farmer's nutrition by allowing them to consume their own processed fruit during and off seasons.

7.2 Technological fruits and vegetable processes

Fruits and vegetable undergo some technological processes with respective to the type of purpose to which it is to be put to use. Fruits and vegetables are either processed for direct (table) eating or for further process into juice, or other products. The followings are few processes a product can undergo.

Controlled ripening

A number of climacteric fruits, notably bananas, mangoes, papaya, pears, tomatoes, and avocados are picked relatively green and are subsequently ripened. Traditional rural methods of ripening bananas and mangoes involve putting the fruit into a pit in the ground or into a heap on the surface and covering with a tarpaulin or branches. Home ripening is also possible using other extremely low-tech practices such as placing fruits to be ripened in a paper bag with a ripe piece of fruit, close loosely and check in a few days.

Figure 7-1: Ripening banana in paper bags and mango in sack

A simple way to ripen fruits at home in small amounts is to use a ripening bowl. Fruits that require ripening should be placed into the bowl with a ripe apple or ripe banana (or any other high ethylene-generating product). The bowl shown below is made of clear molded plastic and has ventilation holes around the top. Using this method, ripening will take from one to four days.

Figure 7-2: Plastic ripening bowl

For tomatoes, technical grade ethylene gas is introduced into the room at a concentration of about 100 ppm for about 48 hours. Ripening can also be initiated by using ethylene generated by passing ethanol over a bed of activated alumina. This method is safer than using pure ethylene gas.

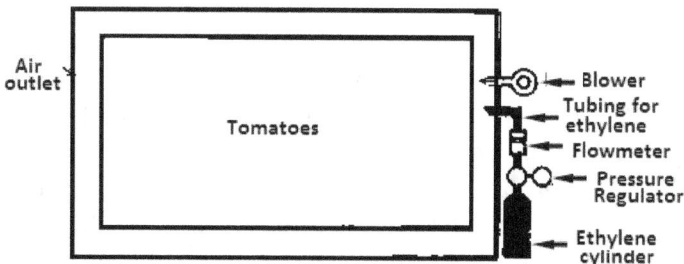

Figure 7-3: Controlled ethylene ripening of tomatoes

Small-scale wholesalers and retailers can ripen fruits in bins or large cartons by placing a small quantity of ethylene-generating produce such as ripe bananas in with the produce to be ripened. Cover the bin or carton with a plastic sheet for 24 hours, and then remove the plastic cover.

Figure 7-4: Controlled ripening of banana

In some cases ripening is induced by placing small sachets of calcium carbide among the fruit, which in the humid atmosphere produces acetylene. Other frequently used techniques include is the exposure of fruit to calcium carbide or smoke produced by burning leaves or wood thereby producing ethylene as one of the products of incomplete combustion. However, product quality may be reduced by this procedure, especially if immature fruit is used.

In modern practice, the fruit are packed in vented cartons stacked on pallets, and fruit temperature is controlled by forced air circulation as in a cooling facility. The fruit to be ripened ideally is placed in an airtight ripening room maintained at a constant temperature (18–21°C for most fruits, but 29–31°C in the case of mango).

Figure 7-5: Controlled ripening rooms

Controlled degreening

Controlled degreening sometimes is carried out on citrus grown in the tropics. Many citrus cultivars are mature before the green colour disappears from the peel. The breakdown of chlorophyll and production of a rich orange colour requires exposure to low temperature during maturation, and this explains why mature citrus frequently is sold green on markets in the humid tropics, where even night temperatures may not drop much below 25°C. Degreening is carried out in ripening rooms. The most rapid degreening occurs at temperatures of 25 to 30°C but the best colour occurs at 15 to 25°C.

Figure 7-6: Degreening of orange

Blanching

Blanching is a process involve dipping vegetables or fruits into boiling water or suspending them in steam for 1.5 to 3 minutes. This process can achieve a number of objectives.

1. First, it kills microorganisms on the surface and in the case of green vegetables destroys the catalase enzyme and inhibits the peroxidase enzyme, both of which can produce off odors and flavors in storage.
2. Second, it removes air from intercellular spaces (assisting in the formation of a head-space vacuum in cans) and softens tissues to facilitate filling containers.
3. Blanching may or may not help to preserve the colour in green vegetables, it is used to crack and loosen the skins of tomatoes, sweet potatoes, and beets prior to peeling.

Fruit is not usually heat blanched because of the damage from the heat and the associated sogginess and juice loss after thawing. Instead, chemicals are commonly used without heat to inactivate the oxidative enzymes or to act as antioxidants. Some produce requires blanching before freezing or drying, and fruits such as apples, pears, peaches and apricots are sometimes treated with sulfur being dried.

Freezing

Freezing is one of the oldest and most widely used methods of food preservation, which allows preservation of taste, texture, and nutritional value in foods better than any other method. The freezing process is a combination of the beneficial effects of low temperatures at which microorganisms cannot grow, chemical reactions are reduced, and cellular metabolic reactions are delayed (Delgado and Sun, 2000).

Figure 7-7: Frozen products

Freezing process: The freezing process mainly consists of thermodynamic and kinetic factors, which can dominate each other at a particular stage in the freezing process. Prior to freezing, vegetables are normally blanched. The material to be frozen first cools down to the temperature at which nucleation starts. Before ice can form, a nucleus, or a seed, is required upon which the crystal can grow; the process of producing this seed is defined as nucleation. Once the first crystal appears in the solution, a phase change occurs from liquid to solid with further crystal growth.

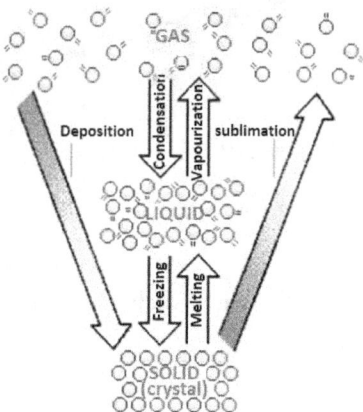

Figure 7-8: A schematic illustration of overall freezing process

Therefore, nucleation serves as the initial process of freezing, and can be considered as the critical step that results in a complete phase change (Sahagian and Goff, 1996). Among several factors affecting freezing time in relation to both the product frozen and freezing equipment, the most important are:

- Dimensions and shape of product, particularly thickness
- Initial and final temperatures
- Temperature of refrigerating medium
- Surface heat transfer coefficient of product
- Change in enthalpy
- Thermal conductivity of product

Freezing is a very popular and effective means of preserving many types of vegetable. Freezing is generally not very effective for cut fruit because the fruit is susceptible to browning and oxidation upon exposure to air. However, some whole fruits such as berries are frozen satisfactorily.

Types of freezing equipment

The industrial equipment for freezing can be categorized in many ways, namely as equipment used for batch or in-line operation, heat transfer systems (air, contact, cryogenic), and product stability. The rate of heat transfer from the freezing medium to the product is important in defining the freezing time of the product. Therefore, the equipment selected for freezing process characterizes the rate of freezing.

Air-blast freezers: The air blast freezer is one the oldest and commonly used freezing equipment due to its temperature stability and versatility for several product types. Freezing is accomplished by placing the food in freezing rooms called sharp freezers.

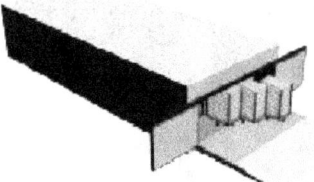

Figure 7-9: Air blast freezer

Tunnel freezers: In tunnel freezers, the products on trays are placed in racks or trolleys and frozen with cold air circulation inside the tunnel. In order to allow air circulation, optimum space is provided between layers of trolley, which can be moved continuously in and out of the freezer manually or by forklift trucks.

Figure 7-10: Trolley in a tunnel freezer

Belt freezers: Belt freezers were first designed to provide continuous product flow with the help of a wire mesh conveyor inside the blast rooms. Modern belt freezers functions by providing a vertical airflow to force air through the product layer.

Figure 7-11: The cross-section of a spiral belt freezer

Fluidized bed freezers: The fluidized bed freezer, a fairly recent modified type of air-blast freezer for particular product types, consists of a bed with a perforated bottom through which cold air is blown vertically upwards (Rahman, 1999). The system relies on forced cold air from beneath the conveyor belt, causing the products to suspend or float in the cold air stream

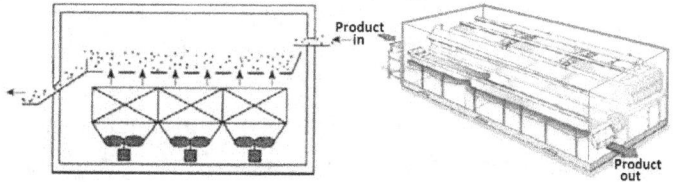

Figure 7-12: Cross-sectional view of a fluidized bed freezer

Contact freezers: Contact freezing is the one of the most efficient ways of freezing in terms of heat transfer mechanism. In the process of freezing, the product can be in direct or indirect contact with the freezing medium. For direct contact freezers, the product being frozen is fully surrounded by the freezing medium, the refrigerant, maximizing the heat transfer efficiency. A schematic illustration is given in Figure 7-13. For indirect contact freezers, the product is indirectly exposed to the freezing medium while in contact with the belt or plate, which is in contact with the freezing medium (Mallett, 1993).

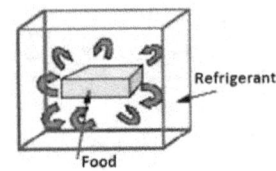
Figure 7-13: Direct contact freezer

Indirect contact freezers: In this type of freezer, materials being frozen are separated from the refrigerant by a conducting material, usually a steel plate. The mechanism of indirect contact freezer is shown in Figure 7-11. Indirect contact freezers generally provide an efficient medium for heat transfer, although the system has some limitations, especially when used for packaged foods due to resistance of package to heat transfer.

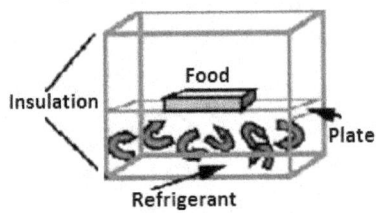
Figure 7-14: Indirect contact freezer

Immersion freezers: The immersion freezer consists of a tank with a cooled freezing media, such as glycol, glycerol, sodium chloride, calcium chloride, and mixtures of salt and sugar. The product is immersed in this solution or sprayed while being conveyed through the freezer, resulting in fast temperature reduction through direct heat exchange

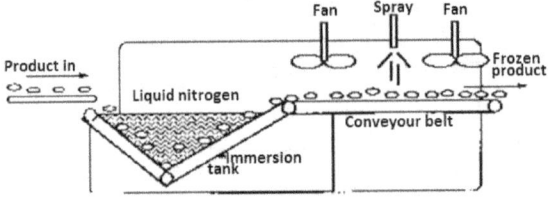
Figure 7-15: Illustration of a typical immersion freezer (Fellows, 2000)

Contact belt freezers: This type of freezer is designed with single-band or double-band for freezing of thin product layers as shown in Figure 7-14. The design can be either straight forward or drum. Typical products frozen in belt freezers are fruit pulps, egg yolk, sauces and soups (Persson and Lohndall, 1993).

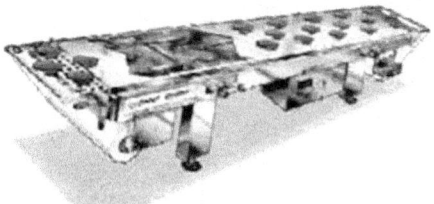
Figure 7-16: Contact belt freezer

Plate freezers: The most common type of contact freezer is the plate freezer. In this case, the product is pressed between hallow metal plates, either horizontally or vertically, with a refrigerant circulating inside the plates. Pressure is applied for good contact as schematically shown in Figure 7-17.

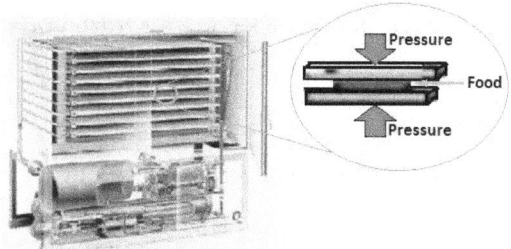

Figure 7-17: Plate freezer with a two-stage compressor and sea water condenser Air

This type of freezing system is only limited to regular-shaped materials or blocks like beef patties or block-shaped packaged products.

Cryogenic freezers: Cryogenic freezing is a relatively new method of freezing in which the food is exposed to an atmosphere below -60 °C through direct contact with liquefied gases such as nitrogen or carbon dioxide.

Frozen product packaging

Proper packaging of frozen food is important to protect the product from contamination and damage while in transit from the manufacturer to the consumer, as well as to preserve food value, flavour, colour, and texture. There are several factors considered in designing a suitable package for a frozen food. The package should be attractive to the consumer, protected from external contamination, and effective in terms of processing, handling, and cost.

Figure 7-18: Packaged frozen products

There are typically three types of packaging used for frozen foods: primary, secondary, and tertiary. The primary package is in direct contact with the food and the food is kept inside the package up to the time of use. Secondary packaging is a form of multiple packaging used to handle packages together for sale. Tertiary packaging is used for bulk transportation of products

Figure 7-19: Types of packaging for frozen products

Freeze-drying

In freeze-drying, the product is first frozen and then placed in a vacuum-tight enclosure and dehydrated under vacuum with careful application of heat, the pressure being kept substantially below 4.6 mm of Hg. The process has a number of advantages which include no reduction in volume, much more of the no aqueous volatile constituents, flavours etc are retained.

Figure 7-20: Dehydrated vs freeze-dried plantain

Freeze-dried foods also rehydrate more rapidly than other dried foods. Although many vegetables can be freeze-dried, one disadvantage is that freeze-dried vegetables are more susceptible to oxidative deterioration than air-dried vegetables.

Drying and sulphuring

Drying is one of the oldest food processing and preservation techniques known to man. The essential feature of the process is to reduce the moisture content to a point at which enzymatic or microbial damage no longer occurs. Fruits (e.g., apricots, peaches) are halved and pitted and then laid out by hand on wooden or plastic trays. Horticultural produce can be dried using direct or indirect solar radiation. The simplest method for solar drying is to lay produce directly upon a Hat black surface and allow the sun and wind to dry the crop. Nuts can be dried effectively in this way.

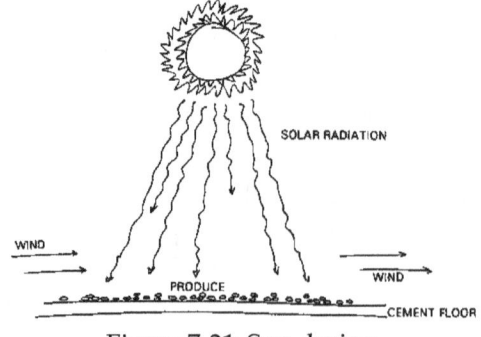

Figure 7-21: Sun drying

Before they are placed in the sun to dry, or put into a high temperature drier, many fruits are first treated with sulphur dioxide in a process known as *sulphuring* to prevent undesirable changes in colour and any additional microbial and enzyme activity, and to retain a residual concentration of 100 ppm in the final product.

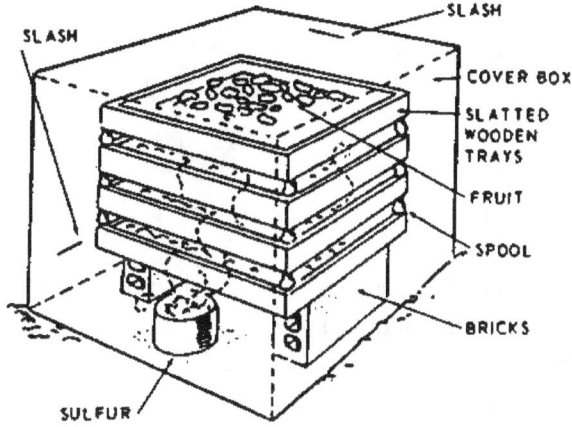

Figure 7-22: Sulphuring box

Low cost sulfuring box can be constructed from a large cardboard box that is slashed in several places to allow adequate ventilation. Trays for drying are stacked using bricks and wooden spools as spacers.

The SO_2 taken up by the fruit displaces the air from the tissue and softens the cell walls so that drying occurs more easily, destroys enzymes that might darken the colour, and actually produces some colour enhancement. To maintain maximum quality, dried fruits should preferably be stored at 10 to 15°C.

Desiccation

Drying or dehydrating vegetables is one of the many procedures advocated for preserving the available perishable food supply. Conservation by desiccation is obvious. Desiccation is the state of extreme dryness, or the process of extreme drying. A desiccant is a hygroscopic (attracts and holds water) substance that induces or sustains such a state in its local vicinity in a moderately sealed container. The desiccation of processed fruits and vegetables is performed in eco-friendly dryers on natural gas.

Figure 7-23: Desiccated fruits

Desiccated or dried vegetables are almost equal to the fresh, and are put up in such a compact and portable form as easily to be transported.

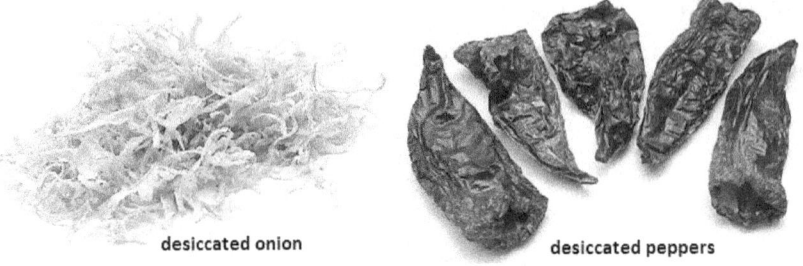

desiccated onion   desiccated peppers

Figure 7-24: Desiccated vegetables

Desiccated coconut is a major product used in the confectionary industry as a bulking material. Fresh coconuts are shelled whole and the outer brown skin or testa pared off with a knife. The coconut water is drained away and the flesh is cut and washed and then sterilized in boiling water or in steam blanchers at 88°C for 5 minutes. The pieces then are shredded into a fine wet meal in a hammer mill and dried to 25% to 30% moisture content in a steam-heated counter flow multistage drier at 77 to 82°C for 40 to 45 minutes. Finally, the product is size-graded and packed.

Processing into purees, pastes, and edible leathers

To prepare these products, fruits such as tomato and mango are first peeled, destoned, and sliced, and the slices pulped in a homogenizer. The pulp then is sterilized by raising the temperature to at least 75 °C (mangoes). Tomato purée is a processed food product, usually consisting of only tomatoes, but can also be found in seasoned form.

Figure 7-25: Processed tomato puree

Tomato purée differs from tomato sauce or tomato paste in consistency and content; tomato purée generally lacks the additives common to a complete tomato sauce, and does not have the thickness of paste. The puree is then canned or sealed in polythene pouches for long-term storage and marketing. Fruit "leathers" can be made by sun drying or forced-air dehydration of a puree layer on a thin plastic film.

Figure 7-26: Processed tomato paste

Processing into flour

A number of vegetable and fruit crops are used to produce flour, two of the major ones being soybean and coconut. Beans are first dried to 13% moisture content, cleaned, and then tempered by storing at 10% to 12% moisture content and 25°C for 7 to 10 days and then dehulled.

Figure 7-27: Processed wheat flour

The dehulled beans then are conditioned by raising the temperature to 70°C, thereby increasing the amount of oil that can be extracted. The conditioned beans then are flaked by passing them through rolls. Oil then is extracted by the solvent process. The flakes then are dried to 10% moisture content, toasted, and ground to a meal.

Aseptic processing and packaging

In contrast to canning, where the hot-filling and post sealing heat treatments are used to produce commercial sterility, in aseptic processing both the product and the packaging are made commercially sterile before the filling and sealing operations, and therefore no post sealing heating is necessary.

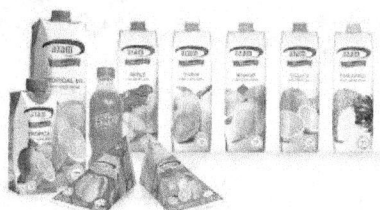

Figure 7-28: Aseptic processing and packaging

The objective of aseptic processing is to obtain a product that is stable for 2 to 3 months, and preferably for 6 months, without refrigeration. Prior to packaging, the liquid or semi liquid product is heated quickly to a temperature at which it is commercially sterile and then cooled. The packages (e.g., coated cartons) are subjected to a combination of chemical and heat treatment. Higher temperatures are more effective in killing off relatively heat-resistant bacteria, while the shorter exposure time results in reduced loss of flavour, colour, and nutrients. This is sometimes known as high-temperature/short-time processing.

Fermentation into alcoholic beverages, vinegar, sauces and other products

Fermentation occurs naturally in the presence of bacteria, yeasts, and molds. Bacteria produce acids, which tend to act as preservatives; yeasts produce alcohol, which is also a preservative. Fruits are fermented to produce alcoholic beverages, including apples (to produce cider), pears (to produce perry), bananas and a wide variety of berry fruits (to produce wines and liqueurs). Traditional alcoholic-beverage preparation relies on the naturally occurring yeasts and sugars to initiate the fermentation process.

Figure 7-29: Alcoholic beverages

Processing into jams and pickles

Four naturally occurring edible preservatives in very common use are sugar, vinegar, salt, and vegetable oil. Jams are made by boiling fruit in sugar syrup until the liquid becomes relatively stiff. Sometimes additional pectin is added to cause the jam to set to a stiffer consistency upon cooling. The jam is poured into clean, sterilized containers while hot, sealed airtight, and then cooled.

Figure 7-30: Fruit jams

In some cases it also is desirable to cover the surface of the jam with a thin layer of wax. Pickling of vegetables and unripe fruit (such as papaya and mango) usually involves some heating with the addition of spices, sugar, and other flavorings, followed by immersion in vinegar (acetic acid) or oil in glass jars.

Canning

The canning process involves cutting, washing, blanching, and then hot-loading into sealed can (which is usually made from tin-plated steel). The can is then held at a prescribed temperature for a given time to ensure that all microorganisms are killed, before being quickly cooled.

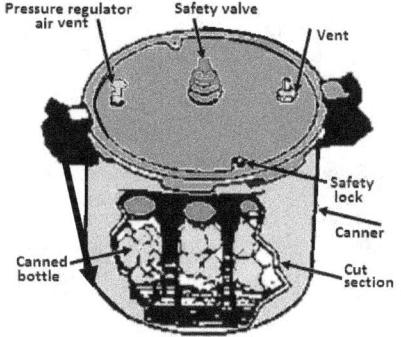

Figure 7-31: Pressurized steam retort

After the can is sealed, it is heat-treated, usually by immersion in boiling water or in a pressurized steam retort. Canning is extremely valuable, because it prevents the reentry of microorganisms completely; the life of the product (even at ambient temperature) can be several years, provided the integrity of the seal is maintained.

Irradiation

Food irradiation is a method of treating food in order to make it safer to eat and have a longer shelf life. This process is not very different from other treatments such as pesticide application, canning, freezing and drying. The end result is that the growth of disease-causing micro organisms or those that cause spoilage are slowed or are eliminated altogether. This makes food safer and also keeps it fresh longer.

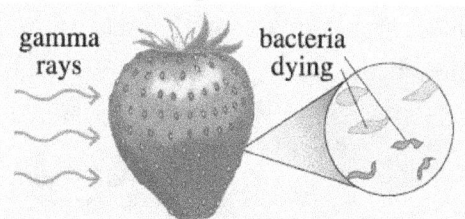

Figure 7-32: Irradiation

Food irradiated by exposing it to the gamma rays of a radioisotope -- one that is widely used is cobalt-60. The energy from the gamma ray passing through the food is enough to destroy many disease-causing bacteria as well as those that cause food to spoil, but is not strong enough to change the quality, flavor or texture of the food.

Ionizing radiation treatment has four potential uses, namely the disinfestations of insect pests, extension of shelf life by retarding ripening and sprouting, inhibition of rotting, and disinfection of material affected by harmful organisms such as salmonella.

Irradiation currently is used mainly for inhibition of sprouting in potatoes and onions. At present Japan makes the widest use of irradiation for these products. Health concerns, in particular the relatively unknown effects of free radicals and other chemical changes produced by the irradiation process, will continue to put limits on the application of irradiation to foods.

7.3 Technological process operations for fruits and vegetables

The typical series of operations in a technological flow process for fruits and vegetables processing into various products are illustrated in Figure 7-33.

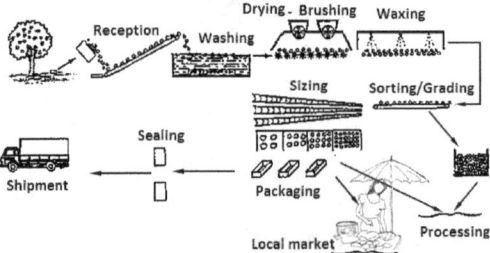

Figure 7-33: Fruit processing operations

The following technological processes have been identified.

7.3.1 Pre-processing procedures

Reception, quality and quantity control

Fresh fruits and vegetables are checked at the reception, then they are carefully sanitized, peeled and washed before being processed.

Figure 7-34: Fruit quality control at reception

Variety of quality controls is needed to identify which fruit belongs to an accepted variety as not all are suitable for different technological processes. The only reliable method for evaluating the quality is the combination of data obtained through taste control; percentage of soluble solid by refractometer, consistency/texture measured with simple penetrometer etc.

Figure 7-35: Fruit quality control at reception

Fruit reception at the processing center is performed mainly for the following purposes:

1. Checking of sanitary and freshness status
2. Control of varieties and fruit wholeness
3. Evaluation of maturity degree.
4. Data collection about quality received in connection to the source supply outside growers/farmers' own farm.

Some useful checks to be performed for raw material control at reception are summarized below

1. *Product quality assurance checks*: Checks are carried out at each delivery/raw material lot for colour, texture, taste, flavour, appearances, refractometric extract, number per kg, variety, sanitary evaluation etc.
2. *Product property checks*: Check for the followings at each ten lots for (the same raw material): density, water content using oven method, total sugars, reducing sugars, total acidity are required to determine the soluble solid contents, titratable acidity etc.
3. *Audit checks*: The type of analysis for audit will be adapted to the specific fruits that are received. Audit check is done at every six months- on five different lots for ascorbic acid, mineral substance, tannic substance, peptic substance etc.

The quality control made at the reception for stored products is to guarantee that the product has been obtained by natural procedures and responds to the quality and packing commercially required and according to the most demanding best international food practices and procedures.

Temporary storage before processing

This step has to be short as much as possible to minimize flavour losses, texture modification, weight losses and other losses dues to deterioration that can take over this period. Some basic rules for this step are as follows

1. Keep product in shades or cool storage
2. Avoid dust as much as possible
3. Avoid excessive heat
4. Avoid any possible contamination
5. Store in a place protected from possible attacks by rodents, insects etc.

Cold storage is always highly preferred to ambient temperature; for this reason, cool room is used for each processing centre.

Fruit packing line

Small scale equipment for packing produce includes a receiving belt, washer and sorting table.

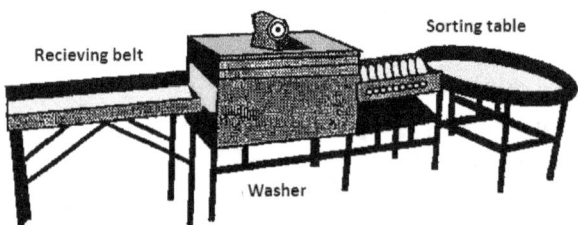

Figure 7-36: Fruit waxing operation

Washing

Harvested fruit is washed to remove soil micro-organisms and pesticide residues. Fruit washing is a mandatory process in order to avoid the pollution of washing tools and or equipment and the contamination of fruit during washing.

Figure 7-37: Fruit washing

Washing of fruit can be carried out by immersion, by spraying or shower or by combination of these two processes which is generally the best solution for pre washing and washing. Some usual practices in fruits washing are:

1. Addition of detergents or 1.5% HCl solution in washing water to remove traces of insect.
2. Use of warm water about 50 °C in the pre-washing phase.
3. Higher water pressure in spray/ shower washers.

Washing must be done before the fruit is cut in order to avoid loss in high nutritive value of soluble substance (vitamins, minerals, sugars etc).

Sorting

Fruit sorting covers two main separates process operations:

a. Removal of damaged fruit and any foreign bodies
b. Qualitative sorting based on organoleptic criteria and maturity.

The most important initial sorting criteria are for variety and maturity. However, for some fruit in special processing technologies, it is advisable to proceed to a manual dimensional sorting (grading). Mechanical sorting for size is usually not done at the preliminary stage.

Figure 7-38: Brushing and hand removal of damaged fruits before grading

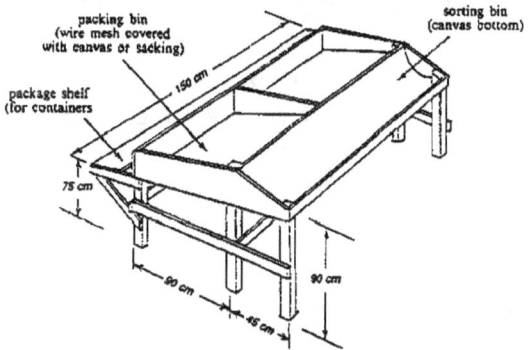

Figure 7-39: Fruit washing bin

Mechanically, three types of conveyors aid in sorting of fruits. The simplest is a belt conveyor, where the sorter handled the produce manually in order to see all sides and inspect for damage. A push-bar conveyor causes the produce to rotate forward as it is pushed past the sorters. A roller conveyor rotates the product backwards as it moves past the sorter.

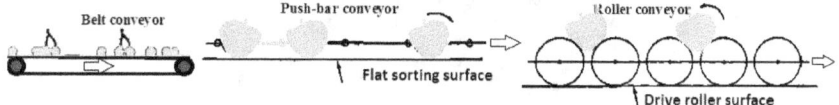

Figure 7-40: Sorter conveyors

Sizing

Round produce units can be graded by using sizing rings. Rings can be fashioned from wood or purchased ready-made in a wide variety of sizes.

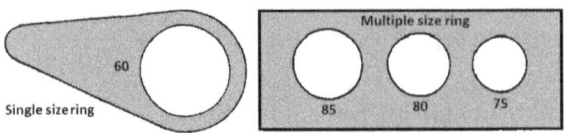

Figure 7-41: Single size hand held sizing ring

The rotary cylinder sizer illustrated below is composed of five hollow cylinders which rotate in a counterclockwise motion when driven by an electric motor. Each cylinder is perforated, with holes large enough to let fruits drop through.

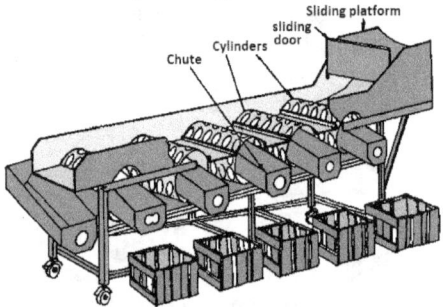

Figure 7-42: Rotary cylinder sizer

The first cylinder has the smallest diameter holes, and the fifth has the largest holes. When fruits fall through, they are caught on a slanted tray (the chute), and roll into the containers as shown. Take care that the distance of the drop is as short as possible to prevent bruising. Oversized fruits are accumulated at the end of the line. This equipment works best with round commodities.

Trimming and peeling (skin removal)

This processing step aims at removing the parts of the fruit which are either not edible or different to digest especially the skin.

Figure 7-43: Rotary cylinder sizer

Up till now the industrial peeling of fruit was performed by three processes.

a. Mechanically
b. By using water steam
c. Chemically by dipping the fruits in a caustic soda-solution at a temperature of 90-100°c: the concentration of this solution as well as the dipping or immersion time varying according to each specific case.

Cutting/slicing

This step is performed according to the specific requirement of the fruit processing technology.

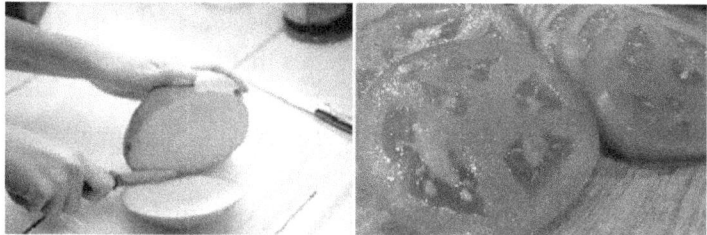

Figure 7-44: Fruit processing operations

Heat blanching

Fruit is not usually heat blanched because of the damage from the heat and the associated sogginess and juice loss after thawing. Instead, chemicals are commonly used without heat to inactivate the oxidative enzymes or to act as antioxidants and they are combined with other treatments.

Chemical treatment

Ascorbic or citric acid dip or vitamin C minimizes fruit oxidation primarily by acting as an antioxidant and itself becoming oxidized in preference to other compounds. Sliced fruits are dipped in sulphur dioxide solution for about 2-3 minutes and then removed. The slice is allowed to stand for about 1-2hrs so that the SO_2 may penetrate the tissue before processing.

Cleaners, sanitizers, and disinfectants

The following list contain a partial list of postharvest chemical treatment agents (cleaners, disinfectants, and sanitizer) allowed for products in order to maintain qualities.

1. *Acetic acid* – allowed as a cleanser or sanitizer. The vinegar used as an ingredient must be from an organic source.

2. *Alcohol (ethyl)* – allowed as a disinfectant. Alcohol must be from an organic source.
3. *Alcohol (isopropyl)* – may be used as a disinfectant under restricted conditions.
4. *Ammonium sanitizers* –Quaternary ammonium salt may be used on non-food-contact surfaces. Its use is prohibited on food contact surfaces, except for specific equipment where alternative sanitizers significantly increase equipment corrosion. Detergent cleaning and rinsing procedures must follow quaternary ammonium application.
5. *Bleach* – calcium hypochlorite, sodium hypochlorite, and chlorine dioxide allowed as sanitizers for water and food contact surfaces.
6. *Detergents* – allowed as equipment cleaners. This category also includes surfactants and wetting agents.
7. *Hydrogen peroxide* – allowed as a water and surface disinfectant.
8. *Ozone* – considered *GRAS* (Generally Regarded as Safe) for produce and equipment disinfection.
9. *Peroxyacetic acid* – water disinfectant and fruit and vegetable surface disinfectant.
10. There are other additional postharvest treatments that may be used on produce:
11. *Carbon dioxide* treatment
12. *Carbon dioxide* treatment is permitted for postharvest use in modified- and controlled-atmosphere storage and packaging. For crops that tolerate treatment with elevated CO_2 (≥15%), suppression of decay and control of insect pests can be achieved.

Fumigants

Fumigants in fruit processing are allowed if materials are in naturally occurring forms (e.g., heat-vaporized acetic acid). Materials must be from a natural source.

Waxing

Wax must not contain any prohibited synthetic substances. Acceptable source of wax include wood-extracted wax. Products that are coated with approved wax must be so indicated on the shipping container.

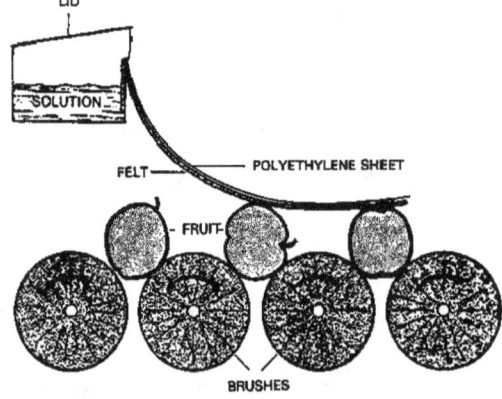

Figure 7-45: Fruit waxing operation

Sugar syrup

Sugar syrup addition is one of the oldest methods of minimizing oxidation by coating the fruit and thereby preventing contact with atmospheric oxygen.

7.3.2 Temporary storage of (fresh) fruit

Some fruits can be stored in fresh state during cold season under some climatic conditions. Fruit for fresh storage have to be harvested before they are fully matured. Sorting and control by quality are mandatory operations. Harvested fruits are transported to storage area as soon as possible. This reduces storage time and accelerates maturation processes. In order to store large qualities of fruit, silos have to be built.

Fruits and vegetables meant for storage must be free from mechanical, damage, insect and disease injury and should be at the proper stage of maturity. Common (unrefrigerated) and cold (refrigerated) storages are the Storage methods generally employed for vegetable and fruits. Common storage lacking precise control of temperature and humidity includes:

a. The use of insulated storage houses
b. Outdoor cellars or mounds'
c. Dugouts

Cold storage allows regulation of temperature and humidity and maintenance of constant conditions by use of a refrigeration and ventilation system. Temperature storage, suitable for very brief storage periods is frequently practiced in the shipping season when large lots are accumulated for load or truck quantities. The refrigerator car or truck is a means of temporary storage while produce is in transit. Short term storage may last for four to six weeks.

Table 7-1: Useful storage life of some food products

Food products	Generalized storage Life (days) at 21°c (70°f)
Animal flesh, fish poultry	1-2
Dried, salted, smoked meat and fish	360and more
Fruits	1-7
Dried fruits	360 and more
Leafy vegetables	1-2
Root crops	1-20

Source: Descrosier (1977)

Postharvest handling is the final stage in the process of producing high quality fresh produce. Being able to maintain a level of freshness from the field to the dinner table presents many challenges. Farmers who can meet these challenges will be able to expand his or her marketing opportunities and be better able to compete in the marketplace.

7.4   Fruit and vegetable processing technologies

A wide range of technology is available for the processing of fruit and vegetables. Various processes (sometimes integrated, sometimes not) yielded one type of fruit or vegetable products Thus, for instance mango can be canned, dried, frozen, or made into puree, sauce, chutney, pickles, concentrate, jam, beverages etc. Papaya can be made to jams, pickles, chutney, nectar, toffee etc., Coconut can be processed to produce oil, milk, cream, desiccated coconut, pie fillers, soft cheese, meal, jam, soap etc. Some of the processing technologies involved in the processing of fruits into finished products include:

1. Drying by dehydration technology
2. Osmo dehydration technology
3. Semi-processed fruit technologies
4. Technology for fruit in syrup "products"
5. Fruit juice processing technology.
6. Processing of fruit bars

7.4.1 Drying and dehydration technology

The sequence of operation employed in the production of dried / dehydrated fruit is presented thus:

Harvesting: Fruit for drying are harvested in a mature condition to obtain maximum yield.

Quantity control: This step is needed in order to maintain finished product standards and to ensure a sound economic operation.

Grading: Maturity and size grading may be needed for individual fruit.

Preparation for drying: Small berry is applied not require preparation cut fruits such as apples as may require washing cutting & covering prior to drying.

Chemical treatment: Alkali dipping is applied to grapes and sometimes to prunes but not to other fruits. Cut fruits are exposed to the fumes of burning sulphur for 2-3 hours. Sulphur dioxide is added to prevent darkening during drying and storage.

Sun drying or artificial drying: Prepared and pretreated fruit is spread out rays and dried in the sun or in factory drying tunnels employing heated air at temperature of 68°c to 74°c.

Bulk storage: The dried fruit is stored in bulk in bins known as "sweat boxes" for 1-9 months until required for packaging. During this time moisture diffuses the wetter fruit pieces to the drier ones and the fruit gradually acquires uniform moisture content.

Sorting and repackaging: The removal of sub standard fruits, loose dirt, stores or other debris is practiced just prior to packaging. The fruit may also be washed at this stage. And if it is too dry, it may be allowed to regain some moisture packing.

7.4.2 Processing of fruit bars

The fruit bar processing involves one single major operation which is drying the fruit pulp after mixing it with suitable ingredients, it can be used to produce mango, banana, guava or mixed fruit bars. A dual-powered processing fryer working by solar energy during the day, by electric or steam at night with cross-flow movement of air and controlled temperature from 55°c at the begging of processing to a highest of 70°C) is well suited for dehydration of the pulp to the desired moisture level of 15-20%.

Figure 7-45: Fruit bars

Mango fruit bars

Fully ripped mangoes are selected and washed in water at room temperature. The peeled fruit is cut into slices and passed through a helicoildal pulper to extract the pulp. The required amount of sugar to adjust Brix (the unit measure for total solids in fruit) of the mixed pulp to 25 degrees Brix is then added.

Table 7-2: Main material qualities to prepare approx. 100kg of fruit bars

Type of fruit	Fruit required in kg	Pulp obtained in kg	Sugar required in kg	Yield (% of fresh fruit) approx.
Mango	720	360	33	14
Banana	600	260	60	17
Guava	406	325	60	20
Mongo+ banana	540+150	360	35	15
Papaya + banana	500 + 140	336	54	23

Source: Amoriggi (1992), FAO (1990).

Citric acid is added to inhibit the micro–organisms during drying. The mixture is then heated for two minute at 80°c and partially cooled. The mixture is transferred to stainless steel trays which have been smeared with glycerin. Each tray is loaded with 12.5kg of mixture. Drying could be carried out by a dual powered dryer for a total of 26hrs; 10hrs by solar energy at 55°c and 16hrs by electric energy or steam power at 70°c.

The moisture content after drying should be between 15 and 20%. Suitable pieces shapes are wrapped in cellophane papers, packed, in cartons and stored at ambient air temperature. The unsuitable shape and size pieces are cut into small variety and used to prepare, along with cashew a variety of "cocktail mixtures".

Banana fruit bars

Banana varieties which give smooth pulp without scrum separation must be used for this purpose. Ripe, suitable are selected. Soak the peeled fruit in 0.3% citric acid solution for about 10minutes (lime or lemon juice can replace citric acid). The drained fruit are pulped to obtain smooth pulp. The rest of the procedure is the same as in the mango bar.

Figure 7-46: Banana fruit bar

7.4.3 Osmotic dehydration in fruit processing

Osmotic dehydration is a useful technique for the concentration of fruit reached by placing the solid food in sugar or salt aqueous solutions of high osmotic pressure. It gives rise to two major simultaneous counter-current flows:

a. Significant water flow out of the food into the solution
b. Transfer of solute from the solution into the food.

Application: The effect of osmotic dehydration as a pre-treatment are mainly related to the improvement of some nutritional, organoleptic and functional properties of the product. Heat damage to colour and flavour is minimal since the dehydration is effective at ambient temperature. Also discolouration is prevented.

The various applications of the technique as a unit operation in the food area are summarized in the flow diagrams below (Figure 7-47) together with the process parameters regarded as optimal.

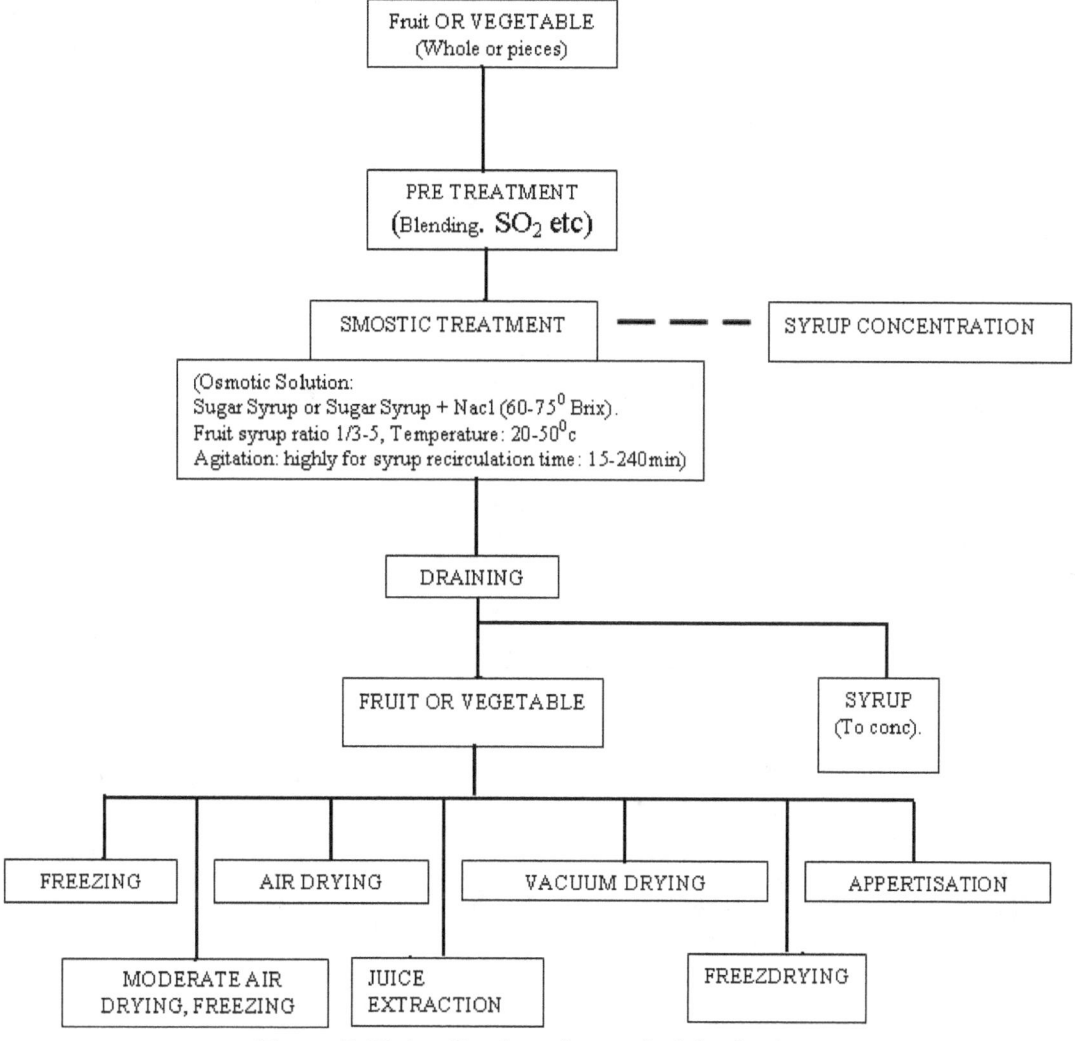

Figure 7-47: Application of osmotic dehydration

Drying: Air drying following osmotic dipping is commonly used in tropical counters for the production of dried food. The combination of osmotic with solar drying has been put forward mainly for tropical fruits. A cycle of 24hrs has been suggested in osmo-dehydration and drying operation. Osmotic dehydration and vacuum drying produced a puffy product with honey comb like texture. Osmotically, dried banana retained move puffiness and a crisper texture then simple vacuum dried ones.

Appertisation: The combination of osmotic dehydration and appertisation improves canned fruit preserves. The feasibility of the process is called osmo-appertisation. The products obtained are stable up to 12 months at ambient temperature.

Freezing: Products obtained in this process are called hydro-frozen and the concentration step is air drying. The commercial feasibility of osmo-dehydration of banana has been studied and the process scheme is shown in flow in Figure 7-48.

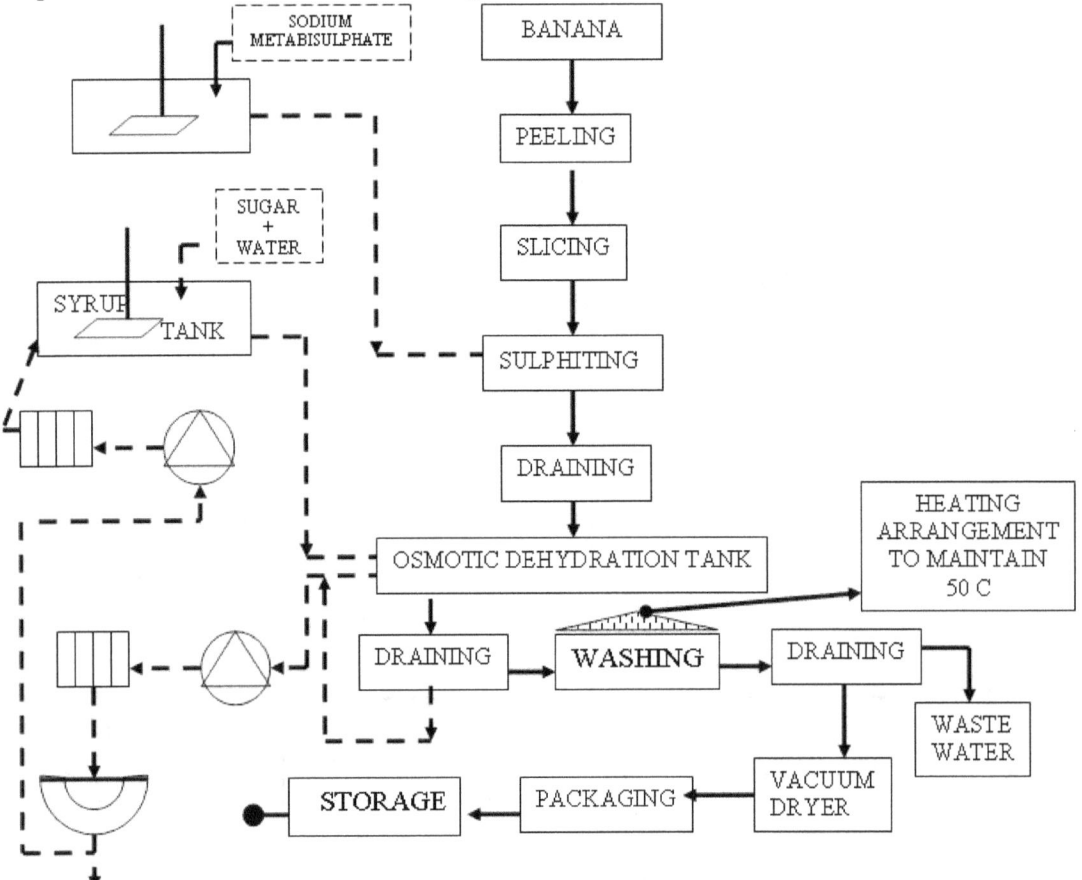

Figure 7-48: Flow diagrams for osmotic dehydration of banana

Handling, sorting, packaging and storage of dried and dehydrated fruits: The characteristics of various products can be assessed either by experience or by instrumentation. When a handful of fruit is squeezed tightly in hand and then released, the individual pieces should drop readily.

When drying is completed, the material should be sorted either on trays or on a table in order to remove poor quality and colour and any foreign matter. Sorted products are packed in polythene bags and packed in cartons or jute bags before, transported.

Deterioration of dried fruit during storage: Dried fruits are considered as a relatively perishable commodity in the same category as cereals, and similarly stored products. They are subject to deterioration resulting from mould growth, insect and mite infestation airs physical chemical changes.

Mould growth: Mould growth occur when the moisture content of dried fruit is allowed to exceed the maximum permissible level for safe storage. The safe moisture level for dried fruit is much higher than those for other similar commodities.

Insect infestation: Insect may begin its infestation in the field before harvest, may continue during bulk storage after drying and unless controlled, may occur in the finished product during storage.

Mite infestation: Severe mite infestations are often associated with the growth of osmophilic yeasts in fermenting dried fruit products. Many of these mites are unable to complete development in the absence of yeast. Such infestations are difficult to eradicate and affect consumer acceptance of the contaminated products.

7.4.4 Technology of semi-processed fruit products

Semi-processed fruit products are manufactured in order to be delivered to country processing centers where they will be further processed in consumer oriented finished products like jams, jellies, syrups, fruit in syrups etc. Preservation of this product is achieved by chemical means, freezing on by pasteurization. The choice of preservation for each individual case is a function of the product type and the shelf life needed.

Technological flow sheet for semi-processed fruit

Sorting: Sorting is needed in order to remove sub standard fruit with would diseases etc and all foreign bodies

Washing: Is obligation in order to remove all impurities which cannot be eliminated at the processing step in the finished product.

Coring and cutting: Mainly for a better utilization of preservation space in receptacles and is not mandatory; this will be defined by customer or supplier agreements standards. This operation is performed by mechanical means. Preservation is carried out with 60% SO_2 solution which is added to the prepared fruit in the quantity needed to obtain the preservation dosage level.

7.4.5 Technology of "fruit in syrup" products

This type of product is represented by fruit (whole, halves pieces) covered by a sugar solution and preserved by pasteurization. In these products, sugar does not have a preservation effect but only sweetening role. Technological flow sheet covers the following steps:

1. Sorting done mechanically but frequently by manual picking on sorting belt.

2. Washing performed in equipment with pan and sprays
3. Clearing removal of leaves etc. skin removal by one of the described methods for some fruits.
4. Cutting is applied only to pomace fruits by mechanical means.
5. Preliminary heat treatment to soften the tissues of hard fruits or enzyme inactivation.
6. Receptacle filling either manually or mechanical (fitting tables etc) then sugar syrup is added and 03% citric acid also added.
7. Preheating/emptying. Air is eliminated from fruit tissues by preheating. It may be done in steam or in water so that syrup temperature reaches 80-90⁰c and is maintained for 10-15 minutes.
8. Cooling should be intensive to avoid discolourations and colour modification.

7.4.6 Technology for fruit juice

Fruit juices are products for direct consumption and are obtained by the extraction of cellular juice from fruit. This operation can be done by pressing on by diffusion. Juice processing cover two finished products categories:

1. Juices without pulp (clarified or not clarified).
2. Juices with pulps (necters)

Natural juices are extracted products obtained from one fruit only,

Figure 7-49: Natural fruit juices

Concentrated juices: This is juice obtained by removal of a major part of product's water content.

Figure 7-50: Concentrated fruit juices

Mixed juices are products obtained from the mixture of two or three juices from different species or by addition of sugar.

Figure 7-51: Mixed fruit juices

Technological processing of juice without pulp

This operation is carried out in centrifugal separators with a speed of 6000 to 6500 rpm. Filtration of clarified juice can be carried out with kieselger and betonite as filtration addictives in press filters.

Figure 7-52: Fruit juice clarifier

Steps for processing of juice without pulp

1. Fruits juices must be prepared from sound and matured fruits only.
2. Washing of the fruit must be thorough and sorting operation carried out on moving inspection belts or sorting tables.
3. Crushing/grinding/disintegration is applied in different ways and depends on fruit type. Crushing for grapes and berries and grinding for apples, pears. Disintegration for tomatoes, mangoes etc. Specific equipments are required for these operations.
4. Enzyme treatment is by adding enzymes at about 50^0c for 30min. extraction yield is improved and better juice colour and product taste. Extraction of fruit juice is achieved thorough pressing.
5. Heating before extraction facilitates pressing colour fining and protein coagulation.

6. Diffusion is an alternative step for juice extraction.
7. Juice clarifying- centrifugation achieves a separation of particles in suspension in the juice.

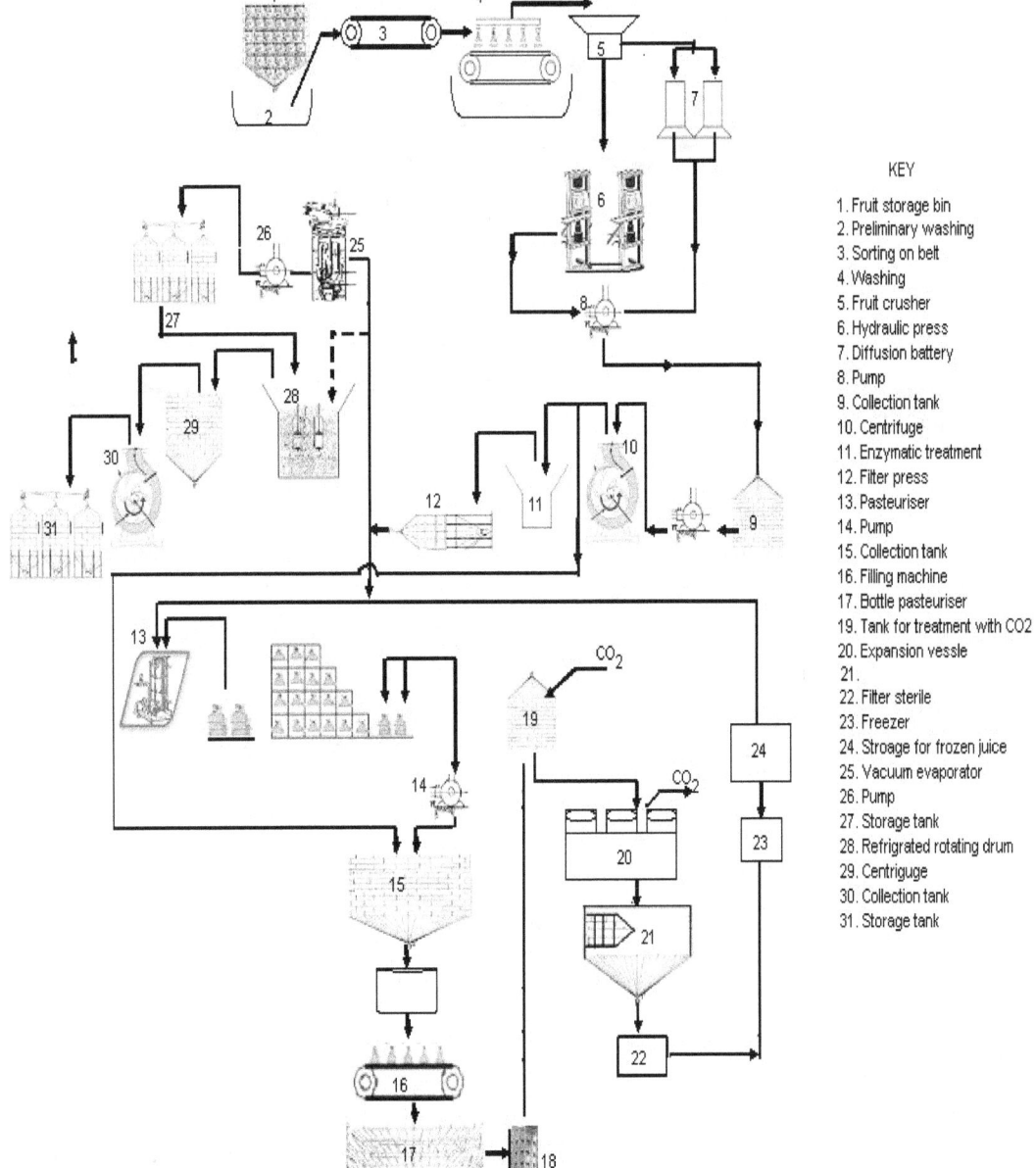

KEY

1. Fruit storage bin
2. Preliminary washing
3. Sorting on belt
4. Washing
5. Fruit crusher
6. Hydraulic press
7. Diffusion battery
8. Pump
9. Collection tank
10. Centrifuge
11. Enzymatic treatment
12. Filter press
13. Pasteuriser
14. Pump
15. Collection tank
16. Filling machine
17. Bottle pasteuriser
19. Tank for treatment with CO2
20. Expansion vessle
21.
22. Filter sterile
23. Freezer
24. Stroage for frozen juice
25. Vacuum evaporator
26. Pump
27. Storage tank
28. Refrigrated rotating drum
29. Centriguge
30. Collection tank
31. Storage tank

Figure 7-52: Technological flow sheet for processing juice without pulp

Technological flow sheet for fruit juices with pulp ("necters"):

This process is divided at industrial scale into two categories of operations:

a. Processing for obtaining Juice
b. Juice conditioning for preservation.

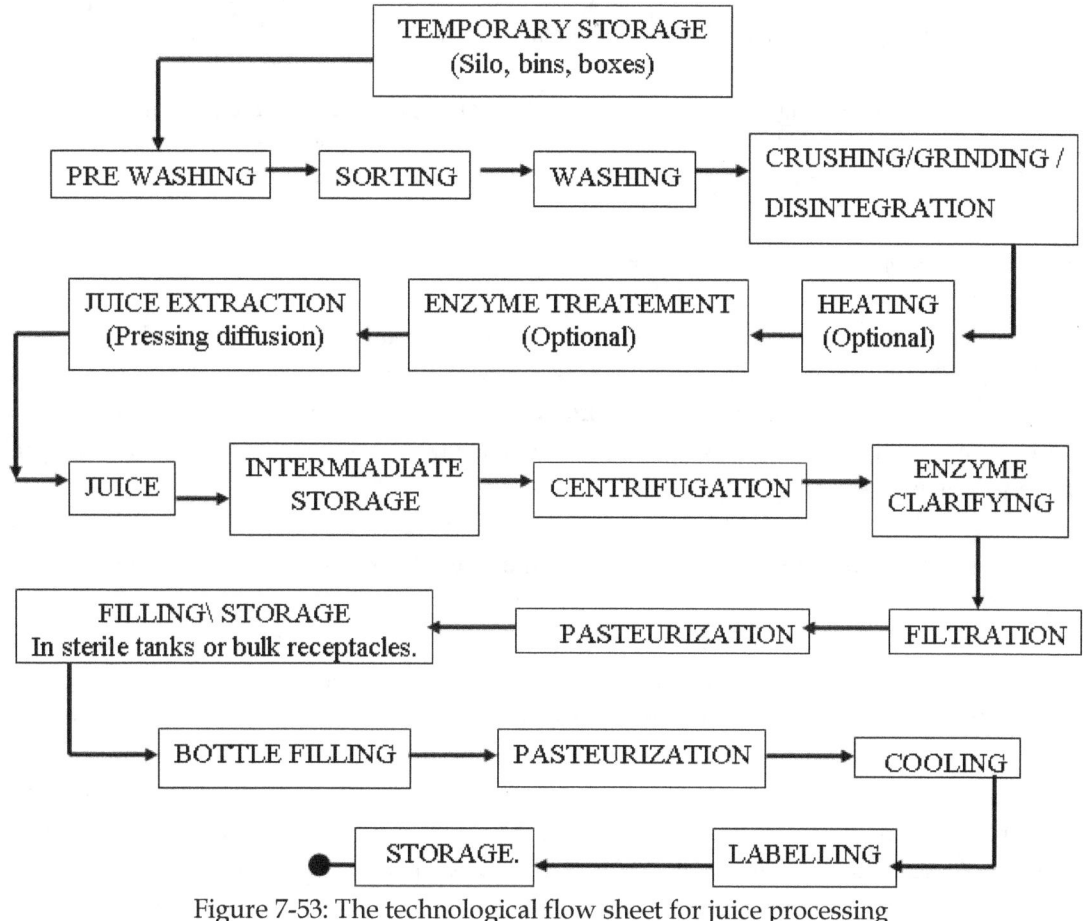

Figure 7-53: The technological flow sheet for juice processing

The general technological flow-sheet for juice processing in represented below in Figure 7-53

7.5   Recent trends in fruit processing

The number and variety of fruit products available to consumers has increases substantially in recent years. The fruit industry has benefited from the importance of these products in a healthy diet. Traditional processing and preservation technological such as freezing and drying together with the move recent commercial introduction of chilling continue to provide the consumer with increased choice. New heating methods are developed (e.g. UHT. Microwave, Ohmic) and freezing methods like the cryogenic technologies combined with new packaging materials and technologies e.g. aseptic, modified atmosphere packaging.

The overall trend in new fruit products is added value thus providing increased convenience to the consumers by having much greater variety of ready prepared fruit products. Fruit juices in confectionery products are now left up to the imagination of the manufacturers.

These products must of course hold up to the standards of flavour integrity and product excellence during the shelf life of these products.

Each new fruit processing center needs a good specific preliminary study in hiding among others raw material availability, harvesting and transport means etc. Excessive atomization of processing equipment does not directly imply a good quality finished product.

New postharvest opportunities and advances

Technologists are poised to make further significant contributions to improving the quality of fresh harvested horticultural products. Further refinements can be expected as we learn more about the interaction of pre harvest factors with responses of products to the postharvest environment. The followings have been identified as leverage for scientific and technological drivers for new postharvest opportunities:

1. *Product uniformity*: Customers, especially supermarket buyers, are requiring increasingly uniform and consistent products with narrowing margins of quality specification.
2. *Modulated storage environment*: Opportunities abound in storage environment modulation to continuously minimize respiration rate and hence deterioration especially now that specific knowledge of product response to endogenous CO_2, $O2$, and C_2H_4 at different times and temperatures after harvest, coupled with the sophisticated temperature- and gas-control systems now available. This can be achieved in both static land-based cool stores and in seagoing containers. Intermittent shock treatments to minimize some physiological disorders, including chilling injury, also may be possible.
3. *Development of active packaging:* Considerable efforts are being made to develop active packaging to overcome problems associated with changing temperatures during storage and transit of modified atmosphere packages.
4. *Development of permeability response packaging films*: Development of films that can increase permeability in response to a chemical signal (e.g., ethanol or acetaldehyde produced during anaerobic respiration) may soon become a reality.
5. *Product surface coatings*: The possibility for creating surface coatings, which reduce water loss but at the same time allow the establishment of appropriate internal atmospheres in fruit but do not allow physiologically "dangerous" levels of gases to develop, is exciting.
6. *Internal quality monitor*: Rapid advances are being made in creating equipment that can monitor internal quality attributes of products non-destructively. Already it is possible to grade fruit into size and colour; electronically using near infrared detectors which measures soluble solids content in fruit along grading lines. The ability to separate products accurately and consistently on the basis of specific internal chemical composition creates the opportunity for providing particular quality attributes (taste or sensory) for discerning niche markets.

7. *Genetic modifications*: Genetic manipulation of plants probably offers the greatest opportunity for making long-term sustainable advances in reducing deterioration rates and extending storage and shelf life of horticultural products.

Molecular biology has allowed scientists to incorporate desirable genes into the genetic makeup of traditional crops. Naturally occurring, long-life mutant tomatoes are now available for commercial production. Molecular technology was used to inhibit cell-wall breakdown and hence slow fruit softening, and hence postharvest decay, as well as producing a more viscous paste that is a great advantage for fruit processors.

Tomatoes have been produced with much reduced ethylene production and hence greatly extended shelf life. Similar lines have been created for apple, peach, and kiwifruit and are being evaluated for other production and quality attributes.

7.6   Introduction to food preservation

Among the oldest methods of preserving food from natural deterioration are drying, refrigeration and fermentation. The causes of food spoilage are growth of micro organisms, an enzyme action, oxidation and dehydration. The form of spoilage to which a food is susceptible depends on its composition, structure, specific microorganisms and storage conditions.

Micro organisms therefore are affected by temperature, moisture, oxygen contamination, available nutrients, degree of contamination with spoilage organism, and the presence or absence of growth inhibitors.

Note: The controls of one or more of these factors usually suffice to inhibit microbial spoilage.

Storage at low temperature prolongs food life by decreasing the respiration rate of fruits and vegetables and by retarding the growth of most spoilage micro organism. For example, the life of many foods may be increased by storage at temperatures below 40 °F (4 °C).

Cold storage is less successful with fruits and vegetable with high water content such as melons, tomatoes, cucumber, banana and pineapples. Since most foods can be stored for many years under appropriate conditions, it is important to establish the condition and storage periods that afford the optimum balance between the cost of storage and the changes in quality of stored products.

BIBLOGRAPHY

List of references for further reading

Acland, J. D. 1971. *East African Crops*. London: United Nations Food and Agriculture Organization/Longman.

Controlled atmosphere http://www.agroripe.com/controlled-atmosphere-storage/

Amos, N. D., L. U. Opara, B. Ponter,C. J. Studman, and G. L.Wall. 1994. Techniques to assist with meeting international standards for the export of fresh produce from less developed countries. Proc. X11 CIGR World Congress, Milan, Italy, vol 2, pp. 1587–1594.

Bello R. S., 2012. Agricultural Machinery & Mechanization. *Pub by Createspace 7290 B. Investment Drive Charl US.* ISBN-13: 978-145-632-876-4. https://www.createspace.com/3497673 June 2012

Bello R. S., 2007. Fundamental Principles of Agricultural Engineering Practice. Pub. Climax Printers #26/30 College Rd. Ogui New Layout, Enugu. ISBN: 978-080-015-8.

Bello R. S., Adegbulugbe T. A. and Odey S. O., 2010. Farm Power and Machinery Operations, Repairs and Maintenance. ISBN: 978-3322254-4-3 Pub. Climax Printers #26/30 College Rd., Enugu Nigeria.

Bencini,M. C., and J. P.Walston, 1991. Post-harvest and processing technologies of African staple foods: A technical compendium. Agricultural Services Bulletin 89. Rome: United Nations Food and Agriculture Organization.

Beth Mitcham, Marita Cantwell, and Adel Kader, 2003. Methods for Determining Quality of Fresh Commodities. Perishables Handling Newsletter Issue No. 85 February 1996, Updated 6/16/03

Bodholt, O., 1985. Construction of cribs for drying and storage of maize. Agricultural Services Bulletin 66. Rome: United Nations Food and Agriculture Organization.

Bubel, M. and Bubel, N. 1979. Root Cellaring: The Simple No-Processing Way to Store Fruits and Vegetables. Emmaus, PA: Rodale Press. 297 pp.

Cantwell, M.I. and R.F. Kasmire. 2002. Postharvest Handling Systems: Underground Vegetables (Roots, Tubers, and Bulbs) p. 435-443. In: A.A. Kader (ed.) Postharvest technology of horticultural crops, University of California. ANR Publication 3311.

Cruz, J. F., and A. Diop., 1989. Agricultural engineering in development: Warehouse technique. Agricultural Services Bulletin 74. Rome: United Nations Food and Agriculture Organization.

Dauthy, M. E., 1995. Fruit and vegetable processing. Agricultural Services Bulletin 119. Rome: United Nations Food and Agriculture Organization.

de Lucia, M., and D. Assennato. 1994. Agricultural engineering in development: Post-harvest operations and management of food grains. Agricultural Services Bulletin 93. Rome: United Nations Food and Agriculture Organization.

Dhamija, O. P., and W. C. K. Hammer, 1990. Manual of food quality control: Food for export. Food and Nutrition Paper 14/6, Rev. 1. Rome: United Nations Food and Agriculture Organization.

Dixie, G. 1989. Horticultural marketing: A resource and training manual for extension officers. Agricultural Services Bulletin 76. Rome: United Nations Food and Agriculture Organization.

Enzafruit. 1996. New Zealand apple quality standards manual. Hastings, New Zealand: New Zealand Apple and Pear Marketing Board (Enzafruit International).

FAO. 1984. Manuals of food quality control: Food inspection. Food and Nutrition Paper 14/5. Rome: United Nations Food and Agriculture Organization.

FAO. 1991. Manual of food quality control: Quality assurance in the food control microbiological laboratory. Food and Nutrition Paper 14/12. Rome: United Nations Food and Agriculture Organization.

FAO. 1993. Manual of food quality control: Quality assurance in the food control chemical laboratory. Food and Nutrition Paper 14/14. Rome: United Nations Food and Agriculture Organization.

FAO. 1985. Standardized designs for grain stores in hot dry climates. Agricultural Services Bulletin 62. Rome: United Nations Food and Agriculture Organization.

FAO. 1984. Agricultural engineering of cold stores in developing countries. Agricultural Service Bulletin 19/2. Rome: United Nations Food and Agriculture Organization.

FAO., 1995. Commodity review and outlook 1994–5. Economic and Social Development series 53. Rome: United Nations Food and Agriculture Organization.

FAO., 1985. Commodity review and outlook 1984–5. Economic and Social Development series. Rome: United Nations Food and Agriculture Organization.

FAO., 1986. *Improvement of Post-harvest Fresh Fruits and Vegetables Handling.* Bangkok: United Nations Food and Agriculture Organization.

FAO., 1989. Utilization of Tropical Foods. Food And Nutrition Bulletin 47 (No. 1 cereals, 2 roots and tubers, 3 trees, 4 beans, 5 oil seeds, 6 spices, 7 fruit, 8 animal products). Rome: United Nations Food and Agriculture Organization.

FAO., 1989. Quality control in fruit and vegetable processing. Food And Nutrition Bulletin 39. Rome: United Nations Food and Agriculture Organization.

Farm Electric Centre. 1989. *Grain Drying, Conditioning and Storage.* Stoneleigh, UK: Electricity Council, Farm Electric Centre.

Grierson, W. 1991. HortTechnology and the developing countries. *Horttechnology* (Oct/Dec):136–137.

İkinci A., 1999. The effect different pruning treatments on yield, quality and carbohydrate accumulation in peach, almond and apricot [Ph.D. thesis], Cukurova University,.

Kader, A. A. 1983. Postharvest quality maintenance of fruits and vegetables in developing countries. In *Postharvest Physiology and Crop Preservation*, ed. M. Lieberman, pp. 520–536. New York: Plenum.

Kader, A. A. 1992. Postharvest technology of horticultural crops. Publication 3311, 2nd ed., University of California, Davis.

Kader, A.A. Fruit maturity, ripening, and quality relationships. *International Symposium Effect of Pre- & Postharvest factors in Fruit Storage* ISHS Acta Horticulturae 485. Url: http://www.actahort.org/ (accessed on 3/10/14)

Kat, J., and A. Diop. 1985. Manual on the establishment operation and management of cereal banks. Agricultural Services Bulletin 64. Rome: United Nations Food and Agriculture Organization.

Lancaster, P. A., and D. G. Coursey. 1984. Traditional post-harvest technology of perishable tropical staples. Agricultural Services Bulletin 59. Rome: United Nations Food and Agriculture Organization.

Li, M., D. C. Slaughter, and J. F. Thompson. 1997. Quality of Kensington mango fruit following combined vapor heat disinfestations and hot water disease control treatments. *Postharvest Biology and Technology* 12:273–284.

Lay-Yee, M., and K. J. Rose. 1994. Quality of Fantasia nectarines following forced air heat treatments for insect disinfestations. *HortScience* 27:1254–1255.

Martin, P. G. 1986. Manuals of food quality control: The food control laboratory. Food and nutrition paper 14/1 Rev.1. Rome: United Nations Food and Agriculture Organization.

McGregor, B. 1987. Tropical Products Handbook. USDA Office of Transportation Agricultural Handbook Number 668.

McKay, S. 1992. Home Storage of Fruits and Vegetables. Northeast Regional Agricultural Engineering Service Publication No. 7

Microsoft® Encarta® Reference Library 2002. © 1993-2001 Microsoft Corporation. All rights reserved.

Mizutani F., T. Kogami, D. G. Moon, R. C. Bhusal, K. L. Rutto, and H. Akiyoshi, 2000. "Effects of summer pruning on the number of apical buds near the trunk in slender-spindle trained apple trees grafted on semi-dwarfing rootstocks," Bulletin of the Experimental Farm College of Agriculture, Ehime University, vol. 22, pp. 1–10,. View at Google Scholar

New Zealand Kiwifruit Marketing Board. 1996. *Kiwifruit Quality Standards Manual.* Auckland, New Zealand: United Nations Food and Agriculture Organization.

Organic Certification, Farm Production Planning, and Marketing, UC ANR Publication 7247

Parmar, C. 1991. Some Himalayan wild fruits worth trial elsewhere. Chronica Horticulturae 31(2):19–20.

Plant Disease Management for Organic Crops, UC ANR Publication 7252

Potato Marketing Board. No date. Control of Environment. Part 2. London: Sutton Bridge Experiment Station, Report No. 6

Rice, R. P., L. W. Rice, and H. D. Tindall. 1987. *Fruit and Vegetable Production in Africa*. London: Macmillan.

Russell, D. C. 1969. Cashew Nut Processing. Agricultural Services Bulletin 6. Rome: United Nations Food and Agriculture Organization.

Seung Koo Lee, 1994. Assoc. Prof., Postharvest Technology Lab., Department of Horticulture, Seoul National University, Suwon 441-744, Korea.

Sode, O. 1990. Agricultural engineering in development: Design and construction guidelines for village stores. Agricultural Services Bulletin 82. Rome: United Nations Food and Agriculture Organization.

Soil Management and Soil Quality for Organic Crops, UC ANR Publication 7248

Soil Fertility Management for Organic Crops, UC ANR Publication 7249

Solhjoo, K. 1994. Food technology and the Bahai Faith. *Food Science and Technology Today*, 8(1):2-3.

Tehrani G. and S. J. Leuty, 1987. "Influence of rootstock and pruning on productivity, growth, and fruit size of European plum cultivars," Journal of the American Society for Horticultural Science, vol. 112, pp. 743–747,. View at Google Scholar

Tracey-White, J. D. 1991. Wholesale markets: Planning and design manual. Agricultural services bulletin 90. Rome: United Nations Food and Agriculture Organization.

Thompson, A. K., 1996. *Postharvest Technology of Fruit and Vegetables*. Oxford: Blackwell.

Trevor Suslow (2000). Postharvest handling for organic crops by the Regents of the University of California, Division of Agriculture and Natural Resources.

Vigneault, C., Raghavan, V.G.S., and Prange, R. 1994. Techniques for controlled atmosphere storage of fruits and vegetables. Research Branch, Agriculture and Agri-Food Canada, Technical Bulletin 1993-18E.

Walker, D.J. 1992. World Food Programme Food Storage Manual. Chatham, UK: Natural Resources Institute.

Weed Management for Organic Crops, UC ANR Publication 7250

Werkhoven, J. 1974. Tea processing: 1974. Agricultural Services Bulletin 26. Rome: United Nations Food and Agriculture Organization.

Wood, B.W., and J. A. Payne. 1991. Pecan: An emerging crop. Chronica Horticulturae 31(2):21–23.

Titles in author's list

Agriculture & mechanization series

- **Farm power and machinery operations**
- Agricultural machinery & mechanization
- Agricultural engineering: principles and practice (Vol 1)
- Agricultural engineering: principles and practice (Vol 2)
- Farm tractor systems: operations and maintenance
- Timeline of agricultural mechanization

Horticultural series

- Horticultural machinery: equipment and safety
- Fruits and vegetable technologies: management options

Workplace safety and machine technology series

- Agricultural machinery hazards & safety practices
- Workplace hazards risks & control
- Workshop technology & practice
- Technical drawing presentation and practice

Students' handbook series

- Study companion
- Path to exam success

Edited works on sustainable agriculture and environment series

- Sustainable agriculture: prospects and challenges
- Sustainable environmental management: issues and projections

More information available @:

http://www.amazon.com/Segun-R.-Bello/e/B008AL6RI0

Notes

www.ingramcontent.com/pod-product-compliance
Lightning Source LLC
Chambersburg PA
CBHW081309170526
45166CB00011B/3458